A Separate Space

Creating a Military Service for Space

MICHAEL SPIRTAS, YOOL KIM, FRANK CAMM, SHIRLEY M. ROSS,
DEBRA KNOPMAN, FORREST E. MORGAN, SEBASTIAN JOON BAE,
M. SCOTT BOND, JOHN S. CROWN, ELAINE SIMMONS

Prepared for the United States Air Force
Approved for public release; distribution unlimited

RAND PROJECT AIR FORCE

For more information on this publication, visit www.rand.org/t/RR4263

Library of Congress Cataloging-in-Publication Data is available for this publication.
ISBN: 978-1-9774-0466-4

Published by the RAND Corporation, Santa Monica, Calif.

© Copyright 2020 RAND Corporation

RAND® is a registered trademark.

Cover: loops7/E+ via Getty Images.

Support RAND

Make a tax-deductible charitable contribution at
www.rand.org/giving/contribute

www.rand.org

Preface

With the passage of the 2020 National Defense Authorization Act, the U.S. government will launch a Space Force. In February 2019, President Donald J. Trump directed the Department of Defense to create the Space Force, as an independent service within the Department of the Air Force, "to ensure unfettered access to, and freedom to operate in, space, and to provide vital capabilities to joint and coalition forces in peacetime and across the spectrum of conflict."[1] The Air Force asked the RAND Corporation to assist in its efforts to plan for the creation of the new service.

RAND provided analytic support to the U.S. Space Force Planning Task Force to inform efforts to build the Space Force. This report provides recommendations on capabilities that our analysis suggests should be included in the Space Force, considers career field sustainability for the fledgling service, and examines other defense organizations to draw lessons for the Space Force. It also considers what it means to be a service as it seeks to illuminate challenges and opportunities the fledgling Space Force will face.

The research reported here was sponsored by Major General Clinton Crosier, Deputy, Deputy Chief of Staff, Strategy, Integration and Requirements, Headquarters U.S. Air Force, and conducted within RAND Project AIR FORCE.

RAND Project AIR FORCE

RAND Project AIR FORCE (PAF), a division of the RAND Corporation, is the U.S. Air Force's federally funded research and development center for studies and analyses. PAF provides the Air Force with independent analyses of policy alternatives affecting the development, employment, combat readiness, and support of current and future air, space, and cyber forces. Research is conducted in four programs: Strategy and Doctrine; Force Modernization and Employment; Manpower, Personnel, and Training; and Resource Management.

Additional information about PAF is available on our website: www.rand.org/paf/

---

[1] White House, *Establishment of the United States Space Force*, Washington, D.C., Space Policy Directive–4, February 19, 2019.

This report documents work originally shared with the U.S. Air Force on September 20, 2019. The draft report, issued on September 11, 2019, was reviewed by formal peer reviewers and U.S. Air Force subject-matter experts.

Contents

Figures and Tables

Figures

Tables

Summary

Issue

The U.S. government has launched a Space Force with the congressional passage and presidential signature of the National Defense Authorization Act for Fiscal Year 2020. On February 19, 2019, President Donald J. Trump signed a directive ordering the Department of Defense to submit a legislative proposal for the establishment of the U.S. Space Force within the Department of the Air Force as a sixth branch of the armed forces, and Congress has now acted to create the force. To prepare for the new service, the Department of Defense and the Department of the Air Force established the U.S. Space Force Planning Task Force on February 22, 2019, to produce an initial work plan and process.[2] The task force approached RAND Project AIR FORCE and requested assistance during this process—in particular to help identify what activities should transfer into a newly established Space Force.

Approach

The RAND Corporation developed a systematic, analytic approach to determine which units to bring into the Space Force, using formal guidance to frame its analysis. First, we asked how a transfer might affect any of the following four organizational attributes: effectiveness, efficiency, independence, and sense of identity. Then, we assessed a set of career fields to consider whether they would be sustainable in the Space Force. We complemented these analyses with an examination of other organizations that the Department of Defense has created to gain insights into potential challenges that the Space Force might face as it stands up and grows into its role.

---

[2] Secretary of Defense, "Memorandum: U.S. Space Force Planning Team," Washington, D.C.: Office of the Secretary of Defense, February 21, 2019; and Secretary of the Air Force, "Memorandum: Establishment of the US Space Force Planning Task Force," Washington, D.C.: U.S. Air Force, February 22, 2019.

Conclusions

Taking into consideration effectiveness and efficiency, independence, and identity, our analysis suggests that the Space Force should consider the following points as it stands up:

- The majority of space activities (operations, training, and acquisition) in the Department of Defense should be moved into the Space Force.[3]
- Headquarters, Space Force, should manage key functions that are essential to the independence and identity of the service (operational concepts and doctrine development; requirements development and advocacy; planning, programming, budgeting, and execution; and legislative liaison and public affairs).
- The proposed consolidation of activities into the new service will likely create new seams between the Space Force and the other services and the organizations it supports. The services will need to retain appropriate organic space expertise and establish liaisons and new relationships with the Space Force to support their unique needs and requirements.

The Space Force will need personnel with skills in space operations, space intelligence, space acquisition, and other science, technology, engineering, and mathematics disciplines relevant to space operations and acquisition. However, there are some major challenges associated with sustaining small career fields and shortfalls in general officer throughput:

- Although some Space Force career fields will be organic to the Space Force, many career fields will be manned by Air Force officers who are on assignment to the Space Force. Some of those Air Force career fields will need to have developed a "space track" to ensure the additional training and development that will be necessary for the Air Force officers who will serve in the Space Force.
- For the foreseeable future, the Space Force will likely need to draw about one-half of its general officers from the Air Force or other services.

The Space Force will also face challenges in achieving effectiveness, efficiency, independence, and identity:

- The use of space capabilities as a force enabler and the current lack of a coherent doctrine of space warfighting could present a challenge to the Space Force's

---

[3] Space intelligence activities that directly support space operations are included as operational activities. Acquisition-related activities are broken down into the following nine areas: oversight and advocacy; requirements and architecture; policies and procedures; research, technology development, and innovation; programs of record; test and evaluation; sustainment; rapid acquisition; and career field management.

effectiveness in its early years and make it more difficult for the service to build a distinctive identity.

- The small size of the Space Force relative to other services is likely to decrease its leverage in the defense community.

Recommendations

We identified several specific areas that the Space Force needs to address to increase its likelihood of developing into a successful organization:

- Define and clarify space warfighting missions.
  - Develop and promulgate a coherent space warfighting theory with which to indoctrinate Space Force personnel and foster the development of a distinctive identity and culture.
  - Develop weapon and support systems to bring substance to theory, making the Space Force a true warfighting service.
- Control resources and public relations.
  - Advocate for a separate total obligation authority for the Space Force.
  - Advocate for the creation of distinct offices for legislative liaison and public affairs.
- Find creative ways to manage career fields and develop senior leaders.
  - In career fields in which substantive space knowledge and skills are required but the career fields are not large enough to be organic to the Space Force, partner with the Air Force to develop "space tracks" to prepare Air Force officers in those career fields who will be serving in the Space Force.
  - Work with Congress to meet general officer requirements.
- Work closely with other services and organizations to define and manage new relationships.
  - Increase space representation at combatant commands.
  - Craft formal agreements to specify and tailor services that the Air Force will provide to the Space Force.
- Establish a formal monitoring and evaluation process to adapt to changes in organizational priorities and external factors.
  - Adopt an adaptive approach to planning.
  - Implement a formal monitoring and evaluation process as soon as possible to track "before" and "after" performance metrics.
  - Establish decision rules for acting when necessary, according to information coming from the monitoring and evaluation process.

Acknowledgments

This project would not have been possible without tremendous support provided by the U.S. Space Force Planning Task Force, led by Major General Clinton E. Crosier. General Crosier; his deputy, Brigadier General Mark A. Baird; and their staff generously provided their time, perspectives, and data with the RAND team. The project team also benefited from conversations with Major General John E. Shaw and his staff at Air Force Space Command. The team also met with Timothy A. Beyland, Robert Corsi, and Lieutenant General (ret.) Robert Elder to learn more from their perspectives as senior Air Force leaders. Within RAND, the project team benefited greatly from the support of RAND Project AIR FORCE Vice President Ted Harshberger, who generously provided his own time and supported the project with sage and practical advice. We are also grateful to Jim Chow, Ray Conley, Obaid Younossi, Paula Thornhill, and David Orletsky, who helped the team work through and around numerous obstacles. Nicholas Martin, Mary Kate Adgie, Brandon Corbin, Alexandra Evans, and Lisa Wagner conducted analysis and wrote case studies that appear as appendixes to the report. The team was also well served by thorough, thoughtful, and helpful internal reviews of our work from Cynthia Cook, Michael Decker, and Michael Hansen. In addition, Tim Bonds, Wade Markel, Gary McLeod, Brian Persons, Jim Quinlivan, and Bonnie Triezenberg also provided valuable input.

Abbreviations

ACC	Air Combat Command
ACTS	Air Corps Tactical School
AETC	Air Education and Training Command
AFGSC	Air Force Global Strike Command
AFRL	Air Force Research Laboratory
AFSC	Air Force Specialty Code
AFSPC	Air Force Space Command
AOC	Air and Space Operations Center
ARL	Army Research Laboratory
ARSTRAT	Army Forces Strategic Command
ASAT	antisatellite
CIA	Central Intelligence Agency
CONOP	concept of operations
CSO	chief of space operations
CYBERCOM	U.S. Cyber Command
DAF	Department of the Air Force
DARPA	Defense Advanced Research Projects Agency
DIA	Defense Intelligence Agency
DoD	Department of Defense
FOC	full operational capability

FY	fiscal year
GHQ	General Headquarters
GRAB	Galactic Radiation and Background
GWOT	global war on terror
ICBM	intercontinental ballistic missile
IOC	initial operating capability
ISR	intelligence, surveillance, and reconnaissance
ISRG	Intelligence, Surveillance, and Reconnaissance Group
JCIDS	Joint Capabilities Integration and Development System
JROC	Joint Requirements Oversight Council
JTAGS	Joint Tactical Ground Station
LAF	Line of the Air Force
MDA	Missile Defense Agency
MFP-11	Major Force Program 11
MOA	memorandum of agreement
MUOS	Mobile User Objective System
NASA	National Aeronautics and Space Administration
NASIC	National Air and Space Intelligence Center
NDAA	National Defense Authorization Act
NIWC	Naval Information Warfare Center
NRL	Naval Research Laboratory
NRO	National Reconnaissance Office
NSA	National Security Agency
NSSI	National Security Space Institute
OG	operations group

OCSO	Office of the Chief of Space Operations
PNT	positioning, navigation, and timing
PPBES	planning, programming, budgeting, and execution system
RCS	Revenue Cutter Service
RSSC	Regional Satellite Communications Support Center
RTS	Reagan Test Site
SAMOS	Satellite and Missile Observation System
SATCOM	satellite communications
SCO	Strategic Capabilities Office
SDA	Space Development Agency
SFPTF	U.S. Space Force Planning Task Force
SOCOM	U.S. Special Operations Command
SOF	special operations forces
SPD-4	Space Policy Directive–4
SSA	space situational awareness
STEM	science, technology, engineering, and mathematics
STRATCOM	U.S. Strategic Command
THAAD	Terminal High Altitude Area Defense
TT&C	telemetry, tracking, and command
UN	United Nations
USAAF	U.S. Army Air Forces
USASMDC	U.S. Army Space and Missile Defense Command
WSOC	Wideband Satellite Communications Operations Center

CHAPTER ONE

Introduction

With the passage of the 2020 National Defense Authorization Act (NDAA), the U.S. government will launch a Space Force.[1] On February 19, 2019, President Donald J. Trump signed a directive ordering the Department of Defense (DoD) to submit a legislative proposal for the establishment of the U.S. Space Force within the Department of the Air Force (DAF) as a sixth branch of the armed forces, and Congress has now acted to create the force.[2]

The passage of this legislation was notable in that it has been 72 years since the United States last created a new service, the Air Force—having established the Army, the Navy, and the Marine Corps in the early years of the nation's history. More recently, DoD has created service-like entities, including U.S. Special Operations Command (SOCOM) and U.S. Cyber Command (CYBERCOM), both of which have Title 10–related responsibilities, but these organizations do not have the same scope or scale as a military service.[3] Therefore, the current effort to develop a Space Force as a separate branch of the U.S. armed services represents a significant step. Creating a service for space offers an opportunity to increase U.S. military capabilities, but undertaking an initiative of this significance brings with it a range of challenges, for which there is limited recent experience to turn to for insights or lessons learned.

The 2020 NDAA calls on the Space Force to provide "freedom of operation for the United States in, from, and to space" and "prompt and sustained space operations" and to "protect the interests of the United States in space; deter aggression in, from, and to space; and conduct space operations."[4] The NDAA established the Office of the Chief of Space Operations (OCSO), with the chief of space operations (CSO) to be appointed by the President and to serve under the authority, direction, and control

---

[1] Pub. L. 116-92, National Defense Authorization Act for Fiscal Year 2020, Subtitle D, United States Space Force; Sections 951–961, December 20, 2019.

[2] White House, *Establishment of the United States Space Force*, Washington, D.C., Space Policy Directive–4, February 19, 2019; Valerie Insinna, "Trump Officially Organizes the Space Force Under the Air Force . . . for Now," *Defense News*, February 19, 2019.

[3] 10 U.S.C.

[4] Pub. L. 116-92, §952, 2019.

of the secretary of the Air Force. The CSO will be a member of the Joint Chiefs of Staff.[5] The NDAA also establishes the Space Force Acquisition Council to oversee the acquisition of space systems and to integrate programs across the national security space community. It also establishes an assistant secretary of defense for space policy to supervise DoD space warfighting policy.[6] The NDAA does not authorize funding beyond that designated in the NDAA itself and does not authorize the creation of any new military billets.[7]

In February 2020, OCSO submitted a report to Congress on preliminary steps to establish the Space Force.[8] The report sets out guiding principles for the organizational design of the service—in particular calling for a "clean sheet" approach to building the organization to harness innovation. The report mentions that plans for Headquarters, Space Force, have been reduced from 1,000 to 800 billets; calls for the establishment of the Space Training and Readiness Command; and suggests that the Space Force will consider an alternative acquisition system to the Joint Capabilities Integration and Development System.

Prior to congressional action to establish the Space Force, DoD and DAF both began to prepare for the new service by establishing the U.S. Space Force Planning Task Force (SFPTF) to produce an initial working plan and process to guide DoD through the creation of the Space Force.[9] The SFPTF, which was working to develop an initial organizational design for the Space Force, approached RAND Project AIR FORCE and requested assistance during this process.

One of the first steps in developing an organizational design is to determine the Space Force's composition, which will be taken from previously existing organizations and crafted into the new service.[10] Much of the new service will come from the Air Force, but it is likely that space-related elements will also need to be transferred from the Army, Navy, Marine Corps, and other DoD organizations. Determining which activities and units should transfer to the new service could get contentious and will be closely scrutinized by the services, DoD and congressional leadership, and many state and local interests. The RAND Corporation developed an objective and transparent

---

[5]   Pub. L. 116-92, §953, 2019.

[6]   Pub. L. 116-92, §955, 2019.

[7]   Pub. L. 116-92, §960, 2019.

[8]   DAF, *Comprehensive Plan for the Organizational Structure of the U.S. Space Force*, Washington, D.C., February 2020.

[9]   Secretary of Defense, "Memorandum: U.S. Space Force Planning Team," Washington, D.C.: Office of the Secretary of Defense, February 21, 2019; and Secretary of the Air Force, "Memorandum: Establishment of the US Space Force Planning Task Force," Washington, D.C.: U.S. Air Force, February 22, 2019. Upon establishment of the Space Force, the SFPTF became the OCSO.

[10]   DoD, *United States Space Force*, Washington, D.C., February 2019a.

analytic framework to help think through which elements (suborganizations, functions, and activities) and which forces should transfer to the Space Force.

Another major issue that the SFPTF has had to consider is workforce management for the Space Force. Although the Space Force would ideally rely on its own personnel and contractor services to perform each activity transferred to its oversight, given the anticipated small size of the Space Force (an estimated initial size of 16,000 personnel),[11] this could be a challenge for many of the skills the Space Force might need. For instance, the small size could limit variability in assignments to fully develop a service member over the course of his or her career or limit leadership opportunities. Depending on the sustainability of the workforce, transferring certain elements to the Space Force might not be viable, or the Space Force might need to work with other parts of DoD to share access to personnel with the skills that it would need. RAND developed an analytic framework to determine whether a career field could be solely sustained within the Space Force.

We complemented these analyses with an examination of other organizations that DoD has created to gain insights into potential challenges that the Space Force might face as it stands up and grows into its role and offer an adaptive planning construct that can guide future Space Force planning and implementation efforts.

To help identify which activities and units should transfer into a newly established Space Force, RAND developed an analytic approach using guidance provided in the following documents to frame the analysis:

- Space Policy Directive–4 (SPD-4)[12]
- the legislative proposal to establish the Space Force[13]
- the Space Force strategic overview.[14]

We also leveraged literature that examined national security space organizational issues and ways of establishing a separate organization responsible for DoD space activities (discussed in Chapter Two).[15]

---

[11] DoD, 2019a.

[12] White House, 2019.

[13] DoD, proposed amendments to U.S. Code to stand up U.S. Space Force, March 2019b. See also Allison DeVito, "Space Force Legislative Proposal," briefing, Washington, D.C.: Office of the Secretary of the Air Force, Deputy General Counsel for Acquisition, February 27, 2019.

[14] DoD, 2019a.

[15] For example, Margaux Hoar, Eileen Chollet, Kory Fierstine, Chao Pan, and Mark Rosen, *Plan for a United States Department of the Space Force: Executive Summary*, Arlington Va.: Center for Naval Analysis, November 2018; DoD, *Final Report on Organizational and Management Structure for the National Security Space Components of the Department of Defense*, Washington, D.C., August 9, 2018; A. Thomas Young, Edward Anderson, Lyle Bien, Ronald R. Fogleman, Keith Hall, Lester Lyles, and Hans Mark, *Leadership, Management, and Organization*

We asked how a transfer might affect any of the four organizational attributes we identified as the most relevant to the vision and mission of the Space Force—effectiveness, efficiency, independence, and sense of identity—to determine whether a space-related activity or unit should transfer. We define these attributes and discuss the logic we applied to determine which activities and units should transfer in Chapter Three and present the results of the transfer analysis in Chapter Four.

We then considered the potential workforces associated with the activities and units that should transfer to the Space Force (per the analysis in Chapter Four) and asked whether they could be effectively managed and sustained in the Space Force. For the transfers of the activities and units to be viable and for the Space Force to function as an independent service, it will have to be able to create and sustain its own career paths and career fields, and it will have to be able to grow and develop its own general officers. We developed an analytic framework (described in Chapter Five) to determine whether a career field could be solely sustained within the Space Force. Given the limited project resources, we applied this framework to a small selection of career fields of high interest to the Space Force. Once DoD identifies units and activities that are candidates for transfer, all the career fields need to be assessed to determine the viability of those transfers.

Our complementary analyses included the following:

- **Case studies of other defense and intelligence organizations:** The U.S. defense and intelligence communities have, in many cases, created new organizations to address policy problems. We looked to a range of other organizations to gain insights into factors that contributed to success and consider potential challenges that the Space Force might face as it stands up and grows into its role (discussed in Chapter Six and Appendix A). We drew on these insights to assess which activities and organizations should transfer to the Space Force and to develop recommendations for the Space Force.

- **Adaptive planning for the Space Force:** We recognize that transfer decisions and the organizational design for the Space Force could change because of a variety of factors (level of resources, the emergence of political barriers, and so on), and the Space Force will also need to grow and evolve as the threat environment and its missions evolve. To support this natural evolution and deal with uncertainties, the Space Force might need to equip itself with the tools to proceed in its implementation in an adaptive and agile manner. To that end, we examined how an adaptive planning approach could be applied to assess how the Space Force is making progress toward its objectives and how course corrections could be made (discussed in Chapter Seven).

for National Security Space: Report to Congress on the Organization and Management of National Security Space, Alexandria, Va.: Institute for Defense Analyses, July 2008.

Before turning to these topics, in the next chapter, we set the context behind the decision to create a new service, consider what it means to be a service, and provide a brief treatment of DoD activities in space.

We conducted this research prior to the passage of the 2020 NDAA and the publication of the OCSO report to Congress. The Space Force that has been established by the NDAA does not exactly match the administration's earlier guidance for the new service. We believe that our findings and the methods we devised to determine which defense activities should transfer into the Space Force and to determine career field sustainability remain valid. However, it is possible that some details of the report are more faithful to the administration's earlier guidance than to the NDAA.

CHAPTER TWO

Contextual Background for the Space Force

The Trump administration's initiative to create a new service was not made in a vacuum. This chapter considers the factors that have influenced the administration's choice. The chapter begins by discussing what it means to be a military service, as opposed to another type of organization. It also provides a brief history of DoD activity in space before turning to some of the events that led to the February 2019 presidential directive.[1]

What Is a Service?

Because a Space Force has been established, it seems advisable to consider what it will mean to have a separate service for space. By establishing a new service, the U.S. government is using organizational means to consolidate military space activities, increase advocacy for military space, and build new capabilities. In addition to or instead of reorganizing, policymakers could reallocate, add, or subtract resources; alter processes; or even just provide rhetorical support for change. Although administration and DoD leaders did provide statements supporting the formation of a Space Force, their primary effort has been organizational, to build a new service.

The preamble of the U.S. Constitution says that one of the main purposes of the founding of the United States is "to provide for the common defence."[2] The new nation needed a military to defend its borders and to protect commerce at sea. These needs were met by establishing the Army to provide frontier security, the Navy to patrol the seas, and the Marine Corps to provide maritime-based infantry.[3] Congress established the Department of War in 1789, which housed both the Army and the Navy until the establishment of the Department of the Navy, in 1798. The Departments of War and the Navy were cabinet-level entities. Over time, each of the services grew from small

---

[1] White House, 2019.

[2] U.S. Const. pmbl.

[3] For more discussion, see Paula G. Thornhill, *Demystifying the American Military: Institutions, Evolution, and Challenges Since 1789*, Annapolis, Md.: Naval Institute Press, 2019, pp. 58–63.

forces with little training and equipment to larger, highly capable forces. In the aftermath of World War II, DAF was established by the National Security Act of 1947. The new service was buoyed by the theory of strategic bombing and the newly prominent role of nuclear weapons. In 1949, an amendment to the National Security Act of 1947 subordinated the Departments of the Navy, Army, and Air Force from "executive level" departments to "military departments" under the newly named Department of Defense.[4]

Formal legal documents offer some insights but do not clearly define what it means to be an independent armed force or service. Title 10 of the U.S. Code provides the basis for much of the service's activities, but it does not provide much help defining either *armed force* or *service*.[5] When referring to Title 10, most members of the defense community refer to the services' role in training, organizing, and equipping forces. This is more useful, but it does not shed light on what it will mean to establish a Space Force separate from the Air Force, other than that a different entity will be responsible for training, organizing, and equipping space forces.

We could seek to define a service by its leadership and those roles. For example, a military department is led by a civilian secretary, and a service is led by a chief of staff, commandant, or chief of naval operations. The secretariat of a military department provides civilian control and budgetary oversight. Services provide military advice, develop doctrine, and define military requirements. Services develop military capabilities, integrate them into their forces, and generate forces to be employed by joint commanders.[6]

Some defense analysts argue that military departments and services are among the most powerful institutions in the U.S. national security enterprise.[7] Military ser-

---

[4]   Pub. L. 80-253, The National Security Act of 1947, July 26, 1947. For more discussion of the establishment of DoD, see Douglas T. Stuart, ed., *Organizing for National Security*, Carlisle, Pa.: Strategic Studies Institute, November 2000, pp. 5–24. The Goldwater-Nichols Act of 1986 further strengthened the civilian leadership of military departments by assigning them authority for weapon acquisition and budget execution (Pub. L. 99-433, Barry Goldwater Department of Defense Reorganization Act of 1986, October 1, 1986). Set up to formalize civilian control of the military, the military departments are led by civilian appointees. The military services sit within the departments: the Army within the Department of the Army, the Navy and Marine Corps in the Department of the Navy, and the Air Force in DAF.

[5]   Title 10 defines *armed forces* as "the Army, Navy, Air Force, Marine Corps, and Coast Guard," and defines *uniformed services* as "the armed forces, the commissioned corps of the National Oceanic and Atmospheric Administration and the commissioned corps of the Public Health Service" (10 U.S.C. §101).

[6]   For more discussion, see Tom Galvin, ed., *Defense Management: Primer for Senior Leaders*, 1st ed., Carlisle, Pa.: U.S. Army War College, 2018, pp. 80–85.

[7]   Carl H. Builder, *The Masks of War: American Military Styles in Strategy and Analysis*, Baltimore, Md.: Johns Hopkins University Press, 1989, p. 3; and Arnold Kanter, *Defense Politics: A Budgetary Perspective*, Chicago: University of Chicago Press, 1979, pp. 16–17. For a more recent discussion, see S. Rebecca Zimmerman, Kimberly Jackson, Natasha Lander, Colin Roberts, Dan Madden, and Rebeca Orie, *Movement and Maneuver: Culture and the Competition for Influence Among the U.S. Military Services*, Santa Monica, Calif.: RAND Corporation, RR-2270-OSD, 2019, pp. 7–9. For discussion of the tension between the Secretary of Defense and the

vices carry with them the core competency for developing forces, as well as concepts of operations (CONOPs) in particular domains.[8] In a comparative study of the military services, Carl Builder details the identities of the Army, Navy, and Air Force. He writes that the Navy's identity centers on "preserving and wielding sea power as the most important and flexible kind of military power," that the Air Force focuses on "ensuring the independence of those who fly . . . [to] use them for . . . both the freedom of flight and the destruction of war," and that the Army's self-identity centers on being "the essential artisans of war, still divided into their traditional combat arms—the infantry, artillery and cavalry (armor)."[9] Interestingly, Builder does not focus on the Marine Corps or the Coast Guard, as neither enters the defense planning arena as an independent institutional actor with a significant voice in the national approach to strategy or military force planning. This is partly because of their institutional position, partly because of their relative size, and, perhaps, partly because of their own institutional personalities or styles.[10]

With the creation of the Space Force inside DAF, one question is whether the new service will be able to forge an independent identity for itself akin to the Army, Navy, and Air Force. Like the Army, Navy, and Air Force, the Space Force will operate in its own distinctive domain. However, as our analysis indicates, its small size, even when compared with the Marine Corps and the Coast Guard, could pose a challenge. In addition, the current focus on space assets as enablers of combat operations instead of as employers of force could also impede the development of an independent identity.

From the history of the U.S. armed forces, and of state-organized militaries in general, it appears that what is unique about them is that they engage in organized violence in the pursuit of national political objectives.[11] This role makes militaries distinct from other organizations. In the United States, military services gain their purpose, or mission, in that they train, organize, and equip forces to use violence against other states and adversaries of the nation.

---

services, and between civilian and military decisionmakers, see David S. C. Chu and Nurith Berstein, "Decisionmaking for Defense," in Stuart E. Johnson, Martin C. Libicki, and Gregory F. Treverton, eds., *New Challenges New Tools for Defense Decisionmaking*, Santa Monica, Calif.: RAND Corporation, MR-1576-RC, 2003, particularly pp. 15–19.

[8] David Ochmanek writes that "concept development and evaluation . . . can only be credibly done within the services" (David Ochmanek, *Improving Force Development Within the Department of Defense: Diagnosis and Potential Prescriptions*, Santa Monica, Calif.: RAND Corporation, PE-302-RC, July 2018, p. 4).

[9] Builder, 1989, pp. 32–33.

[10] Builder, 1989, p. 9.

[11] See Hedley Bull, *The Anarchical Society: A Study of Order in World Politics*, New York: Columbia University Press, 1995, pp. 184–185. For more discussion, see Paula G. Thornhill, *The Crisis Within: America's Military and the Struggle Between the Overseas and Guardian Paradigms*, Santa Monica, Calif.: RAND Corporation, RR-1420-AF, 2016.

Much has changed since the founding of the American republic, and even since the passing of the National Security Act of 1947. For example, the level of personal danger that individuals might experience can be said to have decreased for many in the U.S. armed forces, and the need to deploy to a combat zone for many has decreased, leading one observer to note that these changes have contributed to a crisis that has changed the face of the American military.[12] The argument is that, because of these and other related changes—for example, the role of cyber operations in potentially causing harm to civilians far from a traditional battlefield—a gap is opening between what the U.S. population thinks of its military and what it actually is.

The creation of a Space Force is part of this trend. The Space Force will likely include a large percentage of civilians relative to uniformed personnel, and many of its military service members will not likely directly employ force or have to deploy far from home to support a military operation. In particular, the direct employment of force versus enabling others to use force seems to be an important distinction between what the Space Force will be, initially, and what other branches of the armed forces do. If the Space Force does not directly employ violence against others to advance U.S. national security, and the other services do, the Space Force could face difficulty finding equal footing with other services in internal DoD competition for resources and in external competitions for personnel, prestige, and public (and congressional) support.

Department of Defense Activities in Space over Time

A discussion of the role of the U.S. armed forces in space will provide context from which to analyze the development of the Space Force. Like many developments in the field of engineering, the initial exploitation of space was a child of military necessity, one born of a desire to gain strategic advantage in the cause of national defense. The Eisenhower administration's incentive for putting satellites in orbit was in response to the need for a capability to monitor military developments in the Soviet Union, particularly those involving nuclear forces. Because of such Cold War exigencies, for the first two decades of the space age, the vast majority of satellites, both U.S. and Soviet, were devoted to strategic missions—reconnaissance, surveillance, communications, and environmental monitoring—in support of national decisionmakers and nuclear forces. Given the vital importance of these systems, U.S. space policies were oriented toward ensuring that they would be allowed to operate unhindered by legal challenge and unthreatened by military competition for the control of space. Therefore, Wash-

---

[12] Thornhill, 2016.

ington created NASA to be the public face of the U.S. space program, emphasizing exploration, scientific discovery, and other "peaceful uses" of outer space.[13]

As the United States and the Soviet Union developed their space capabilities, they became ever more closely integrated with nuclear missions, both for strategic warning and for command and control. And as each side's nuclear arsenal grew, many U.S. policymakers worried that attacking the satellites that each side's nuclear forces depended on might be destabilizing. They concluded that allowing satellites to operate as if space were a sanctuary from war was an interest that Washington shared with Moscow.[14]

In the closing decades of the Cold War, however, two trends emerged that called the viability of the sanctuary assumption into question. First, the United States was increasingly using space-based services to support its conventional military operations, raising the question of whether, in any serious conflict, an enemy combatant would allow U.S. satellites to operate unmolested if they gave U.S. forces advantages on the battlefield. Second, space services came to support an ever-growing proportion of civil and commercial functions in American society. By the end of the Cold War, satellites, once almost exclusively in the purview of military and intelligence operations, were providing essential support for storm tracking and weather prediction and a rapidly expanding array of services in broadcast and global point-to-point communications. This caused some military and civilian leaders to worry about how seriously the U.S. economy or other aspects of the American way of life might be affected if those services were interrupted.[15]

With the end of the Cold War, the trend toward ever-greater military dependence on space-based services accelerated. Space systems remained central to national strategic warning and the command and control of nuclear forces. More significantly, space became a critical linchpin in U.S. conventional military operations. This became evident during the 1991 Persian Gulf War, where, for the first time, national strategic assets were tasked to support conventional combat operations on a large scale. Reconnaissance satellites were retasked from strategic missions, and their products were made more widely available to theater commanders and their staffs. Data from strategic sur-

---

[13] The Eisenhower administration sought a United Nations (UN) ruling supporting the principle of "freedom of space"—i.e., that satellites could legally overfly other countries without violating their sovereignty. However, in 1958, the UN Ad Hoc Committee for the Peaceful Uses of Outer Space rendered an opinion that overflight was legal only if the satellites were being used for "peaceful purposes." From then on, all presidential administrations up to Ronald Reagan's blocked U.S. space weaponization programs, declared that all U.S. space programs were for peaceful purposes, and used NASA as a "stalking horse" to hide U.S. satellite reconnaissance and other military space programs. See R. Cargill Hall, *Military Space and National Policy: Record and Interpretation*, Arlington, Va.: George C. Marshall Institute, 2006.

[14] For a detailed history of U.S. and Soviet developments from the dawn of the space age to the closing decade of the Cold War, see Walter A. MacDougall, *The Heavens and the Earth: A Political History of the Space Age*, Baltimore, Md.: Johns Hopkins University Press, 1985.

[15] Jeffrey L. Caton, "Joint Warfare and Military Dependence on Space," *Joint Forces Quarterly*, Winter 1995–1996.

veillance satellites were brought to bear to provide warning of Scud missile attacks and support coalition efforts to hunt down and destroy the launchers.[16]

Military communications satellites, originally designed to support command and control of nuclear forces, were moved and reconfigured to provide an unprecedented volume of bandwidth to conventional war fighters. And when that capacity was fully subscribed, DoD leased channels on commercial satellites to provide even more.[17] Most significantly, GPS, with an initial operational capability provided by 16 of the planned 21 satellites on orbit, provided coalition ground forces sufficient navigational support to enable the now famous "Hail Mary" maneuver, the flanking movement across open desert that ultimately routed Iraq's frontline forces.[18]

The Gulf War was a threshold event in U.S. military space operations. Having discovered the vast potential of space to support conventional military operations in future conflicts, the U.S. defense community began concerted efforts to modify existing space systems and develop new ones to support tactical users. In the years that followed, a wide array of innovative technologies was developed employing data from space-based assets. In July 1995, GPS achieved full operational capability, with 24 satellites on orbit providing continuous precise navigational support to military and civilian users everywhere on the globe. With that capability, a whole new class of precision weapons emerged, from gravity bombs to cruise missiles, using GPS data to guide them to their targets, and military vehicles on land, in the air, and at sea began using GPS for navigation.[19] New generations of military communications satellites were developed to support conventional operations, and DoD leased ever-greater volumes of bandwidth from commercial communications satellites to meet the increasing demands of U.S. conventional forces.

As these capabilities became available, U.S. aerospace developers began applying a "system-of-systems" concept, networking military sensors, command-and-control systems, and weapons to construct "sensor-to-shooter" linkages, resulting in "network-centric warfare."[20] Space capabilities are crucial in U.S. network-centric operations: Targets are often identified using space-based intelligence, surveillance, and recon-

---

[16] Curtis Peebles, *High Frontier: The United States Air Force and the Military Space Program*, Washington, D.C.: Air Force History and Museums Program, 1997, p. 74.

[17] Thomas A. Keaney and Eliot A. Cohen, *Revolution in Warfare? Air Power in the Persian Gulf War*, Annapolis, Md.: Naval Institute Press, 1995, pp. 242–243.

[18] Hall, 2006, p. 16. In 1991, plans for the fully operational GPS envisioned a 21-satellite constellation and three operational spares. The constellation's "fully operational" status was later redefined as 24 satellites. There are now about 30 operational satellites in orbit.

[19] Although the 1991 Gulf War was famous for the use of precision-guided weapons, none was guided by space-based assets. These weapons were all guided by terrestrial systems, such as laser designation or electro-optical, radio control.

[20] Arthur K. Cebrowski and John J. Garstka, "Network-Centric Warfare: Its Origin and Future," *Proceedings*, Vol. 124, January 1998.

naissance (ISR) sensors; precision weapons are guided to their targets using GPS; and network components are linked via a satellite-borne communications backbone.

These developments have enabled new warfighting capabilities, such as long-endurance unmanned aerial vehicles (UAVs), which are now extensively used for ISR and strike operations in permissive environments. Another space-reliant capability is blue-force tracking systems for better battlespace situational awareness. Both UAVs and blue-force tracking rely on space-based communications systems. As a result of these developments, the U.S. military's conventional warfighting capabilities have become increasingly fast and lethal, despite the rapid drawdown in conventional forces that occurred in the 1990s and continued at a lesser pace for more than a decade afterward. However, there is a downside to such developments. Orbital assets are difficult to defend, and, without space support, the effectiveness of U.S. military forces would be badly degraded.[21]

Although the United States was the first nation to test satellite countermeasures,[22] potential adversaries have noted the vulnerability of (and U.S. dependence on) space assets and are developing capabilities to exploit the situation. For instance, China has embarked on an ambitious program to develop offensive space control capabilities.[23] In January 2007, it destroyed one of its expired weather satellites in a successful test of a direct-ascent, antisatellite (ASAT) missile, the SC-19, launched from a ground-based, mobile transporter erector launcher.[24] The SC-19 is now an operational weapon system, able to engage targets in low earth orbit. And U.S. satellites in low earth orbit are not the only ones threatened by Chinese space weapons. China now appears to be developing direct-ascent ASAT capabilities able to reach targets in medium earth orbit

---

[21] Defense Intelligence Agency, *Challenges to Security in Space*, Washington, D.C., January 2019, p. 11.

[22] In 1959, the United States launched a ballistic missile from an Air Force B-47 bomber at an Explorer VI satellite. From the early 1960s to the mid-1980s, the United States and the Soviet Union pursued counterspace weapons, with the final test in 1985, when a missile launched from an Air Force F-15 fighter destroyed a U.S. solar observation satellite. See Justin Paul George, "History of Anti-Satellite Weapons: US Tested 1st ASAT Missile 60 Years Ago," *The Week*, March 27, 2019.

[23] DoD defines *offensive space control* as consisting of "offensive operations conducted for space negation, where negation involves measures deceive, disrupt, degrade, deny, or destroy space systems or services. Adversaries, both state and non-state actors, will exploit the availability of space-based capabilities to support their operations. In keeping with the principles of joint operations, this makes it incumbent on the US to deny adversaries the ability to utilize space capabilities and services. [Offensive space control] actions targeting an enemy's space-related capabilities and forces could employ reversible and/or nonreversible means." Joint Chiefs of Staff, *Space Operations*, Washington, D.C., Joint Publication 3-14, April 10, 2018a, p. II-2.

[24] This act provoked a storm of protests from the international community, because it littered a band of low earth orbit with debris, endangering all other space assets that pass through it. Chinese authorities learned from the experience—they have tested the SC-19 at least two times since then, but always against ballistic targets, leaving no debris in orbit. See Brian Weeden, *Through a Glass, Darkly: Chinese, American, and Russian Anti-Satellite Testing in Space*, Washington, D.C.: Secure World Foundation, March 17, 2014, pp. 1–2.

and geostationary earth orbit.[25] Moreover, China has developed a wide assortment of directed-energy weapons, including radio frequency jammers to block satellite communications (SATCOM) and low- and high-power lasers to dazzle or damage optical sensors.[26]

Russia is also threatening U.S. space capabilities, although its space control programs do not appear to be as far along as China's. Moscow first began developing such capabilities during the Cold War, but for the first decade and a half after the collapse of the Soviet Union, Russian leaders showed little interest in them. In 2009, however, then–Chief of the Russian General Staff General Nikolai Makarov gave a press conference in which he delivered a detailed report on plans for Russian military reform. Among the many goals he laid out, Makarov said that Russia would need to develop ASAT weapons to destroy U.S. and Chinese satellites.[27] Russia is now working to develop these capabilities.

In March 2018, Russia conducted the sixth test of the Nudol, a direct-ascent ASAT system.[28] First tested in late 2015, the Nudol is part of a suite of next-generation kinetic interceptor systems that Russia is developing, which also includes the S-500/55R6M surface-to-air missile system and the A-235 next-generation antiballistic missile system.[29] Like China, Russia is developing radio communications jammers and directed-energy weapon technologies to dazzle or damage sensitive space-based optical sensors.[30] Of particular concern, both countries have launched experimental satellites that conduct sophisticated on-orbit activities that could be used in offensive space control missions.[31]

---

[25] Todd Harrison, Kaitlyn Johnson, and Thomas G. Roberts, *Space Threat Assessment 2018*, Washington, D.C.: Center for Strategic and International Studies, April 2018, p. 8.

[26] Harrison, Johnson, and Roberts, 2018, pp. 8–9. *Dazzling* is a term used to describe temporarily blinding a satellite's optical sensors by illuminating them with a low-power laser.

[27] Roger N. McDermott, *Russia's Strategic Mobility: Supporting Hard Power to 2020?* Stockholm: Swedish Defence Research Agency, April 2013.

[28] Russia might have conducted another successful test of the Nudol system on June 4, 2019, but international observers are uncertain whether the system tested was the Nudol or the S-500 Prometheus long-range, surface-to-air antiaircraft system. See Ankit Panda, "Russia Conducts New Test of 'Nudol' Anti-Satellite System," *The Diplomat*, April 2, 2018; "S-500 or A-235? Russia Tests New Missile Defence System with Extreme Range," *Military Watch*, June 4, 2019.

[29] Panda, 2018.

[30] Harrison, Johnson, and Roberts, 2018, p. 15; Brian Weeden and Victoria Samson, eds., *Global Counterspace Capabilities: An Open Source Assessment*, Washington, D.C.: Secure World Foundation, April 2018, pp. 2-1–2-31.

[31] Weeden and Samson, 2018, pp. 1-2–1-10, 2-2–2-11. Additional, potential threats to U.S. constellations include attack via high-powered microwaves, cyber intrusion (data intercept or seizure of satellite control), and high-altitude nuclear detonation. Finally, in the terrestrial domain, U.S. space capabilities are vulnerable to kinetic strike on, or sabotage of, the satellite ground stations that serve as telecommunications hubs. See Todd Harrison, Kaitlyn Johnson, and Thomas G. Roberts, *Space Threat Assessment 2019*, Washington, D.C.: Center for Strategic and International Studies, April 2019, pp. 3–7.

China's and Russia's offensive space control programs have alarmed U.S. military leaders, raising questions about whether the United States should resume efforts to develop such weapons.[32] Openly doing so would change the prevailing tenor of U.S. space policy, which has emphasized the peaceful uses of outer space throughout most of its history. The rising threat to U.S. space capabilities presented by China's and Russia's space control programs has also prompted U.S. leaders to call for the formation of an independent U.S. Space Force.

Toward a Presidential Directive

The drive to establish a separate service for space did not begin in 2019, with an initiative from the Trump administration. Awareness of the contribution of space to warfighting grew considerably following the 1991 Gulf War, as did arguments for the establishment of a separate service for space.[33] Several congressionally appointed commissions have examined national security organization for space. In 2001, the Commission to Assess United States National Security Space Management and Organization called for establishing the Presidential Space Advisory Group to provide high-level advice to the President to develop a long-term strategy on space, as well as the Senior Interagency Group for Space in the National Security Council to coordinate space activities across the U.S. government.[34]

In 2008, a congressionally mandated report on U.S. defense space organization found that leadership for military space was "fragmented and unfocused" and called for the development of a core group of government professionals to support space acquisition.[35] In 2018, another congressionally mandated report issued by the deputy secretary of defense called for the establishment of the Space Development Agency to develop and field new capabilities and a proposed U.S. Space Command to develop concepts of operation, doctrine, and tactics, techniques, and procedures.[36]

Advocates for developing a Space Force believed that it would help consolidate military space functions currently spread across the intelligence and defense communities, help develop a cadre of national security space professionals, and create a needed

---

[32] For example, see Lara Seligman, "Space Force Is Trump's Answer to New Russian and Chinese Weapons," *Foreign Policy*, August 10, 2019.

[33] Michael C. Whittington, *A Separate Space Force: An 80-Year-Old-Argument*, Montgomery, Ala.: Air War College, Maxwell Paper No. 20, May 2000.

[34] Commission to Assess United States National Security Space Management and Organization, *Report of the Commission to Assess United States National Security Space Management and Organization*, Washington, D.C., 2001.

[35] Young et al., 2008.

[36] DoD, 2018.

advocate for developing military space capabilities.[37] For example, Congressman Mike Rogers played a leading role in Congress advocating for a Space Corps that would transition to a separate force.[38]

Debates over improving U.S. military capabilities in space featured several organizational choices. DoD could establish a new policy directorate, a new development and acquisition agency, a new combatant command, a Space Corps or Space Force within DAF, or a completely independent Space Force.[39] Of these, the administration pursued a Space Force within DAF. The choice of pursuing an organizational approach was rooted in part in an attempt to minimize the resources expended on the new force. Policymakers have explicitly sought to limit the resources involved in building the Space Force to $2 billion and to limit the number of personnel in the force to 16,000.[40]

The administration issued three documents that guided the effort to establish a Space Force: SPD-4, a legislative proposal, and a strategic overview.[41] SPD-4 calls for a Space Force within DAF but notes that this is a first step that should lead to a future Department of the Space Force.[42] The directive notes that the force should enable "offensive and defensive space operations, and joint operations in all domains." It also notes that establishing a Space Force will allow for the consolidation of existing military forces but directs that NASA, the National Oceanic and Atmospheric Administration, the National Reconnaissance Office (NRO), and other nonmilitary U.S. government organizations not be included in the new force.[43]

Because this approach will result in having two services within DAF, we use the following terminology in the remainder of the report to distinguish between the department and the services within it. When we refer to *DAF*, we refer to the entire

---

[37] For example, see Todd Harrison, "A Space Force Is Worth the Price," *The Hill*, November 27, 2018b.

[38] Phillip Swarts, "Rogers Calls for Separate 'Space Corps' Within the Air Force," *Space News*, April 4, 2017.

[39] Todd Harrison lays out three options: a Space Corps inside DAF and two options for an independent Department of the Space Force (Todd Harrison, *How Much Will the Space Force Cost?* Washington, D.C.: Center for Strategic and International Studies, November 2018a). For information on other options, see Jason Coleman, Adam Talaber, and Matthew Woodward, *The Personnel Requirements and Costs of New Military Space Organizations*, Washington, D.C.: Congressional Budget Office, May 2019.

[40] Mike Gruss and Aaron Mehta, "Space Force to Cost $2 Billion, Include 15,000 Personnel in First Five Years," *Defense News*, March 1, 2019. To be clear, the $2 billion would be applied to the costs of transitioning to the Space Force. The service's total budget has yet to be determined and will depend in part on how many units, assets, and processes are transferred into the service. For more detail, see Kathleen J. McInnis and Stephen M. McCall, *"Space Force" and Related DoD Proposals: Issues for Congress*, Washington, D.C.: Congressional Research Service, April 8, 2019. Since passage of the NDAA, end strength for the Space Force will likely be 16,000 (Pub. L. 116-92, 2019).

[41] White House, 2019; DoD, 2019b; DoD, 2019a.

[42] White House, 2019, p. 1.

[43] White House, 2019, pp. 2–3.

Air Force, including both the secretariat and Air Force as a service, and soon likely to include the Space Force as a service. When we refer to the *Air Force*, we refer only to the Air Force as a service and not as a department. Similarly, when we refer to the *Space Force*, we refer to the entity as a service and not a department.

As directed by SPD-4, the administration also submitted a legislative proposal to amend Title 10 to allow for the creation of the new branch.[44] The legislative proposal also called for creating a chief of staff and a vice chief of staff of the Space Force, both with the rank of general, and for a space staff to support the service.[45]

The strategic overview provided more detail on the administration's thinking on the strategic context underpinning the need for a new service, noting that China and Russia have developed counterspace capabilities that could put U.S. operations at risk.[46] It also provided detailed guidance for how the Space Force should be established. For example, it directed the Space Force to "leverage existing [Air Force] infrastructure, except in performing those functions that are unique to the space domain or that are central to the independence of the new Military Service." The document also laid out a phased transition plan for the service to be established from fiscal years (FYs) 2020 to 2024.[47] According to the overview, during this period, military capabilities from other services would be transferred to the Space Force, along with 15,000 personnel.[48]

---

[44] White House, 2019; DoD, 2019b.

[45] DoD, 2019b, pp. 3–7. The NDAA establishes the CSO instead of a chief of staff of the Space Force

[46] DoD, 2019a, pp. 4–5.

[47] DoD, 2019a, pp. 9–11.

[48] DoD, 2019a, pp. 9–10. Since passage of the NDAA, end strength for the Space Force will likely be 16,000 (Pub. L. 116-92, 2019).

CHAPTER THREE

Planning for the Force: Analytic Approach for Determining Which Activities Should Transfer to the Space Force

This chapter describes an analytic approach that RAND developed to determine which activities and units should transfer to the Space Force. The analytic approach consists of four steps, shown in Figure 3.1.

First, we identified organize, train, and equip activities that the Space Force will need to carry out and then identified relevant DoD units and organizations that are performing those activities.[1] Second, we developed four organizational performance metrics, which we applied to assess whether a particular activity should transfer. Third, we developed question sets tailored to each activity type or category to qualitatively assess the potential effects of transfer on the performance metrics. Fourth, on the basis of the assessment of potential effects on the performance metrics, we applied a logic to determine whether an activity and the associated units or organizations should transfer. This chapter describes each step in detail.

Figure 3.1
Analytic Steps to Assess Which Activities and Units Should Transfer to the Space Force

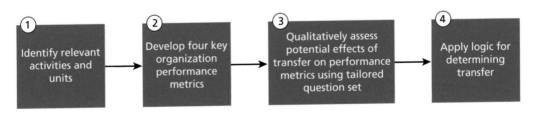

---

[1] The strategic overview lists the following missions as those assigned to the Space Force: space situational awareness; satellite operations and global, integrated command and control of military space forces; global and theater military space operations to enable joint campaigns (including missile warning); space support to land, air, naval, and cyber forces; spacelift and space range operations; space-based nuclear detonation detection; and offensive and defensive space operations to achieve space superiority (DoD, 2019a).

Step 1. Identify Relevant Activities and Units for Analysis: Elements of a Space Force

As a new military service with Title 10 responsibilities, a new Space Force will have the responsibility for organize, train, and equip activities that could then provide operational capabilities to U.S. Space Command and combatant commanders to employ. DoD Directive 5100.01 defines these activities.[2] For the purpose of our analysis, we group them into four functional areas:

1. **operational and training activities** for missions assigned to the Space Force, including space control; space situational awareness (SSA); positioning, navigation, and timing (PNT); ISR; SATCOM; missile warning; space-based nuclear detonation; environmental monitoring; space launch and range operations; and satellite operations[3]
2. **acquisition activities** that develop, procure, and sustain materiel and information systems
3. **major headquarters activities** that coordinate the activities above, including creation and oversight of policy and doctrine; planning, programming, and budgeting; management of external relationships; and provision of enterprise-wide shared services and generic headquarters back-shop support activities
4. **support activities** that recruit and sustain military personnel; operate bases; and provide other support services, including nonoperational communications, comptroller, contracting, logistics readiness, security force, and other support activities.

Operational mission activities will, by definition, be closest to the core mission activities of the Space Force. Training activities are key support activities that the Space Force needs in order to deliver operational capabilities. We grouped the training activities with operational activities for the purpose of our analysis because of the close linkages between the nature of operations and training activities. Acquisition activities are another set of key support activities required to deliver operational capabilities. Per the legislative proposal, the acquisition authority for Space Force systems should remain with the secretariat in DAF.[4] The acquisition-related activities we assessed are those that the Air Force, other services, and DoD organizations currently perform and what Space Force might perform in the future.

---

[2]   DoD, *Functions of the Department of Defense and Its Major Components*, Washington, D.C., Department of Defense Directive 5100.01, December 21, 2010.

[3]   The space missions assigned to the Space Force are described in terms of the space missions listed in Joint Publication 3-14 (Joint Chiefs of Staff, 2018a). For definitions and descriptions of these missions, refer to this document.

[4]   DoD, 2019b.

Training and acquisition are support activities that are integral to the immediate effectiveness of the capabilities of the operational mission activities and involve resources that are specific to these operational missions. The headquarters of the Space Force will provide forces to joint force commanders to perform operational activities, and the Space Force will also provide the activities that support these forces. We group all the other support activities that do not directly support operational missions or do not involve resources that are specific to operational missions. Some of these support activities might be more generic, and some might not require any customization to support space missions.

We identified which units or organizations in various parts of DoD are associated with these activities. Because of the compressed project timeline, we narrowed the scope as follows: We reviewed the active components in the military services and omitted the reserve and guard units. That said, the assessment approach we describe in this chapter could be applied to those units as well. Further, we focused on training required for performing and attaining proficiency in space operations and did not review training for other activities that directly support space operations (e.g., training for space intelligence). Additionally, we did not review organizations that provide individual space education, such as the service academies, Air Force Institute of Technology, and Naval Postgraduate School.

Step 2. Develop Organizational Performance Metrics: Defining Success for the Space Force

Central to any major reorganization, whether in government or the private sector, is clarity among leaders and stakeholders about the strategic objectives driving the new organization. Defining what success looks like for the new Space Force is particularly challenging because of its uniqueness and the rapidly changing threat and technology environment in which it will operate. There are robust academic and business management literatures on the topic of organizational performance metrics aimed primarily at the private sector for which profit is the most important objective.[5] Some of

---

[5] For examples, see Robert B. Carton and Charles W. Hofer, *Measuring Organizational Performance: Metrics for Entrepreneurship and Strategic Management Research*, Cheltenham, UK: Edward Elgar Publishing, 2006; Marc J. Epstein, *Making Sustainability Work: Best Practices in Managing and Measuring Corporate Social, Environmental and Economic Impacts*, Routledge: London, 2008; Pierre J. Richard, Timothy M. Devinney, George S. Yip, and Gerry Johnson, "Measuring Organizational Performance: Towards Methodological Best Practice," *Journal of Management*, Vol. 35, No. 3, 2009; Michael J. Malboussin, "The True Measures of Success," *Harvard Business Review*, October 2012; David Parmenter, *Key Performance Indicators: Developing, Implementing, and Using Winning KPIs*, New York: Wiley & Sons, 2015; B. Aquilani, C. Silvestri, A. Ruggieri, and C. Gatti, "A Systematic Literature Review on Total Quality Management Critical Success Factors and the Identification of New Avenues of Research," *TQM Journal*, Vol. 29, No. 1, 2017; and Clive Keyte, *How to Develop Meaningful Key Performance Indicators: A Step by Step Guide*, Winchester, UK: Intrafocus Limited, 2018.

these approaches are also flexible enough to accommodate performance metrics for the public sector in which multiple public objectives typically are in play.[6] Although quantitative metrics are preferred, the goals and organizational attributes intended for the Space Force do not fit neatly into a quantitative framing beyond those concerning costs.

Regardless of the particulars of different frameworks for constructing organizational performance metrics, most sources emphasize the proposition that a few top-level metrics are vastly preferable to a lengthy list that fails to communicate priorities or provide guidance for decisionmaking. These key metrics need to be specific and measurable, even in qualitative terms. By having a small number of metrics, an organization's leaders can focus on the strategic objectives that matter most to performance and achievement of the mission.[7] We identified the following four organizational attributes that collectively embody the vision and mission of the Space Force as articulated in SPD-4 and the strategic overview:[8]

- **effectiveness** of the joint force, within which the Space Force will operate
- **efficiency** of the use of resources in the joint force, within which the Space Force will operate
- **independence** of the Space Force
- **identity** of the Space Force.

Thus, we use them as top-level organizational performance metrics that should guide the development of the Space Force and decisions about transfers.[9]

Although these four organizational attributes are not explicitly framed as performance metrics in the formal guidance documents, we inferred that they are desired attributes for the Space Force. For instance, the strategic overview states that "estab-

---

[6] For example, the Balanced Scorecard sets up a simple visual approach to illuminating trade-offs across competing objectives, helping decisionmakers focus on the measures of performance that matter most to them (Robert S. Kaplan and David P. Norton, "The Balanced Scorecard: Measures That Drive Performance," *Harvard Business Review*, January–February 1992; Robert S. Kaplan and David P. Norton, *The Balanced Scorecard: Translating Strategy into Action*, Boston: Harvard Business School Press, 1996; Robert S. Kaplan and David P. Norton, *The Strategy-Focused Organization: How Balanced Scorecard Companies Thrive in the New Business Environment*, Boston: Harvard Business School Press, 2000).

[7] Parmenter, 2015, pp. 19–20.

[8] White House, 2019; DoD, 2019a.

[9] We discussed these metrics and other candidate metrics with SFPTF and RAND subject-matter experts. Most of the other candidate metrics were either a subset or a close cousin of the four metrics we selected. For example, *unity of effort* was considered, but we deemed it to be a subset of effectiveness. *Advocacy* was also considered but it is a close cousin of independence. *Warfighting ethos* was another one, but it reflected the higher-level metric of identity. One high-level metric we considered that warranted a deeper consideration was *sustainability*. However, rather than including it as an organizational performance metric to guide the transfer decisions, we chose to treat it as a measure of feasibility of transfer. Thus, we conducted the workforce sustainability analysis as a separate analysis.

lishing a Military Service with dedicated military leadership will unify, focus, and accelerate the development of space doctrine, capabilities, and expertise to outpace future threats; institutionalize advocacy of space priorities to provide for the common defense in all domains; and further build space warfighting culture."[10] The first objective points to effectiveness, the second to independence, and the third to identity. The strategic overview further describes the mission of the Space Force as organize, train, and equip U.S. military space forces to "provide for freedom to operate in, from, and to the space domain; to provide independent military options for national leadership; and to enable the lethality and effectiveness of the joint force." Again, the themes of independence and effectiveness clearly appear as key objectives. Lastly, SPD-4 defines the scope of the establishment of the Space Force as consolidating "existing forces and authorities for military space activities, as appropriate, to minimize duplication of effort and eliminate bureaucratic inefficiencies," indicating that efficiency needs to be considered in the establishment of the new service.[11]

As we will describe in more detail in a subsequent section, on step 3 of our analytic approach, we assessed how a transfer of particular activity to the Space Force might affect these performance metrics to determine whether it should transfer. We offer a brief definition of these four organization performance metrics and discuss examples of how they could be measured qualitatively.

Effectiveness

Effectiveness is a measure of military capability conditional on the available human capital, technological, and funding resources. We define it as the level of military capability that the joint force can achieve within a fixed level of resources. Activities that are organizationally closer to operational missions (for example, space control or satellite operations) can affect the performance of the joint force more than those that do not directly affect space operations (for example, base operations or security). Activities that produce assets customized for use in space operations (for example, unit training or acquisition) are more likely to affect effectiveness than those that produce relatively generic outputs (for example, legal services or contracting).

In general, transferring any activity to the Space Force changes the location of *seams* in the value chains that ultimately yield capabilities for the joint force. For example, advocates often argue that the Space Force can help integrate efforts currently dispersed among various armed services that currently conduct activities in space, allowing these activities to operate more in sync with one another and thereby improving the capabilities of the joint force as a whole.[12] But moving activities to the Space Force

---

[10] DoD, 2019a.

[11] White House, 2019.

[12] For example, see Cristina T. Chaplain, "Space Acquisitions: DoD Continues to Face Challenges of Delayed Delivery of Critical Space Capabilities and Fragmented Leadership," testimony before the Subcommittee on Stra-

also creates new seams between the activities conducted in space and the military operations they support on the land, in the air, and in the sea. These new seams, for example, could slow operations in a kill chain that relies on sensors or communication assets in space to support kinetic attack on the land or in the sea, thereby reducing the capabilities of the joint force.

The scope of our study did not permit us to conduct detailed assessments of such net effects on joint capabilities, which are likely to be heavily context dependent. Rather, we apply the professional judgment of the subject-matter experts on the project team to assess potential effects. More-detailed analysis could more finely identify arrangements for coordinating activities inside and outside the Space Force to achieve the best joint outcomes over the longer term.

Efficiency

Efficiency is the means by which the ends of effectiveness are achieved with the lowest consumption of resources. Cost-effectiveness is a close cousin of efficiency and captures the desirable property of achieving the most effectiveness for a given level of costs. Lesson 3 of Paul Light's top ten lessons on organizational design is the observation that "high performance is not always efficient."[13] The inherent tension between operational performance (effectiveness) and efficiency will need to be carefully analyzed and acknowledged when considering metrics for the Space Force,.

Holding any level of military capability constant, efficiency increases as the level of resources needed to achieve that level of capability falls. For example, the cost of performing activities that display scale economies is likely to rise if they move from a large armed service, such as the Air Force, to the small armed service that the Space Force is expected to be. Scale economies come in many varieties. Examples include the following:

- A significant portion of the cost of headquarters activities is fixed. If the Space Force must recreate headquarters already present in the secretariat, the level of headquarters costs per dollar of activity in the Space Force as a whole is likely to rise. Similarly, if the secretariat must duplicate activities that oversee the Space Force and Air Force, the level of headquarters costs per dollar of activity in DAF as a whole is likely to rise.
- Development and application of common standards tend to better improve cost-effectiveness as the size of the organization in which they are applied grows. Part of this economy comes from spreading the cost of creating and managing stan-

---

tegic Forces, Committee on Armed Services, U.S. Senate, May 17, 2017.

[13] Paul C. Light, *The Four Pillars of High Performance: How Robust Organizations Achieve Extraordinary Results*, New York: McGraw-Hill, 2005, pp. 71–72.

dards across a larger organization. Part comes from detecting best practices and quickly diffusing them across all activities covered by a standard.

- Activities with network economies can generate greater value in larger organizations. For example, the more activities a communications network includes. the more valuable it is to all participants.
- Effective category management yields greater savings in the purchase of generic inputs in larger organizations. Larger organizations can also typically exploit greater leverage in their input markets than smaller organizations.
- Larger organizations are typically more resilient to discrete surprises that degrade their performance and cost. This presumes that the organization can quickly move people, equipment, materiel, and other resources among activities to respond to surprises. This argument covers both the basic insurance rationale for managing surprises—a portfolio of unrelated threats yields a smaller threat to the portfolio as a whole as the size of the portfolio grows—and the mechanical ability to match available supplies of resources to the threat-induced demands for them across an organization as surprises arise.

The relative importance of such scale economies differs across the activities that might transfer to the Space Force. But viewed broadly, all of them point to the likelihood that a smaller armed service, such as the Space Force, is likely to face higher total activity costs than a larger organization conducting the same activities. This indicates that the Space Force needs to be able to justify these higher costs by allowing the joint force to perform more effectively.

Independence and a Sense of Identity

Independence relates to the ability of Space Force leadership to put forth its priorities within the DoD and joint force structure without subordinating those priorities to those of the Air Force or other components. Space Force leaders might not get what they ask for, but they can demonstrate the independence of the force by making their case directly to the Secretary of Defense and to Congress. *Identity* relates to the ability of the Space Force to foster a culture separate and distinct from the Air Force and other services.[14]

Independence and a sense of identity in the Space Force lie at the heart of its ability to increase the effectiveness of the joint force. These elements reflect critical

---

[14] A recent RAND report focuses on diagnostic metrics associated with organizational culture, elements highly relevant to one of the motivations behind establishing a Space Force—which is to develop a distinct organizational culture apart from the Air Force and other services. The report highlights several categories of organizational attributes: culture, strategy, structure and governance, skills, human capital, and accountability. Questions are presented to diagnose the status of these attributes (Kate Cox, Stephen Jolly, Simon Van Der Staaij, and Christian Van Stolk, *Understanding the Drivers of Organisational Capacity*, Santa Monica, Calif.: RAND Corporation, RR-2189-SAATCHI, 2018; see Table 2.2 for a list of performance-oriented questions associated with these categories).

organizational attributes that will likely enable the Space Force to achieve its mission. Advocates of the creation of the Space Force typically argue that the leadership of DAF has been unable to appreciate the full opportunities to operate effectively in space.[15] As space has become more contested, these opportunities have become even more important.

If the Space Force does not have the independence to develop and advocate an alternative approach to operate in space, it might not be able to advance new opportunities as it operates in the bureaucratic politics of DoD. And, as discussed in Appendix A, the Marine Corps and, more recently, the Air Force have found a strong sense of identity to be critical to their ability to stand apart—independently—for a long enough time to create distinct approaches to warfighting that have ultimately improved the capabilities of the joint force.

We view independence and a sense of identity as distinct but difficult to separate. As we argue in more detail in Chapter Four, independence would likely allow the Space Force to create a new doctrine of warfighting, advocate it to external audiences, and instantiate it in the Space Force's plans, programs, and budget. Independence would allow the Space Force to choose the criteria for promoting its own personnel and deciding whom to promote. A sense of identity is the cultural glue that would likely hold together the members of the Space Force long enough to realize a new concept of warfighting and translate the concept into a self-sustaining capability. Independence sets the stage for a sense of identity, which in turn sustains independence over time. We believe that these performance elements are best assessed together and focus more on their mutual, rather than their separate, contributions to outcomes for the joint force.

Step 3. Qualitatively Assess Potential Effects of Transfer on Performance Metrics

We qualitatively assessed the potential effects of transfer on the four organizational attributes by asking a set of questions tailored to each functional area. These questions were focused on key factors that contribute to effectiveness, efficiency, independence, and identity in the context of each functional area. We identified these factors and developed the questions using guidance in SPD-4, the legislative proposal, and the strategic overview;[16] insights from our case studies and discussions with SFPTF members; and a workshop with RAND subject-matter experts held in Pentagon City, Virginia, in May 2019. The tailored questions for each functional area are described in Chapter Four.

---

[15] Douglas Loverro, "Why the United States Needs a Space Force," *SpaceNews*, June 25, 2018.

[16] White House, 2019; DoD, 2019b; DoD, 2019a.

Answers to these questions informed our judgment about the potential effects of a transfer on the four organizational attributes. The project team answered these questions using descriptions of space units and of organizations available in public domain (e.g., fact sheets) and internal expertise on space operations and training.

Step 4. Apply Logic to Determine Which Activities Should Transfer to the Space Force

Using the assessment of potential effects on the four organizational attributes, we applied the following logic to determine whether a particular activity should transfer to the Space Force (see Figure 3.2). If the transfer of any activity increased one of these elements without reducing any of the others, we supported a transfer. If the transfer increased one and reduced another, we applied the professional judgment of the subject-matter experts on the project team to compare the competing effects. If, on net, a transfer appeared to improve the performance of DoD as a whole, we nominated this activity for transfer. If not, we recommended against transfer. Where a transfer from the current service could degrade effectiveness or efficiency, we recommended agreements that the Space Force and current services might pursue to improve effectiveness

Figure 3.2
Logic for Determining Whether an Activity Should Transfer to the Space Force

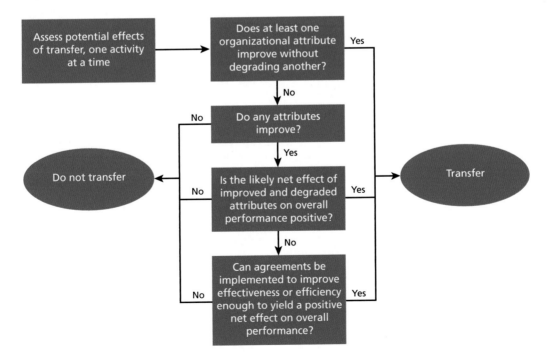

and efficiency while preserving as much of the independence of the Space Force as possible.

Given the limited time allotted for this assessment, we did not conduct interviews with or obtain additional information from personnel who are assigned to these units. We recommend a more detailed review of individual activity and units to arrive at a more refined view of what to transfer and how to manage ongoing relationships between the Space Force and other services. Appropriate decisions about what to transfer and how to manage activities that do not transfer are also likely to change as the Space Force matures and a more refined understanding of its role within DoD emerges.

In the next chapter, we discuss the results of our analysis. We describe the activities and organizations that should transfer to the Space Force according to our analysis and agreements that the Space Force might need to pursue with other services to enhance performance elements.

CHAPTER FOUR

Planning for the Force: Activities and Organizations That Should Transfer to the Space Force

This chapter describes our assessment of which activities and organizations should transfer to the Space Force. We present the assessment in four major sections that correspond to the four functional areas that we described in Chapter Three—operational and training activities, acquisition activities, major headquarters activities, and support activities. Each section describes the relevant activities and organizations included in each functional area and a set of questions tailored to each functional area to assess potential effects of transferring the activities and organizations on the organizational attributes (effectiveness, efficiency, independence, and identity). Each section concludes with the findings of our assessment.

Operational and Training Activities

We define *space operational units* as those that perform or directly support

- operations of a space system, which comprises the space segment (spacecraft and payload), ground control segment (satellite command and control and satellite control networks), and the network segment of a space system (satellite downlink stations and processing, exploitation, and dissemination stations)
- operations of non-space-based systems in support of SSA or space control
- operations of space launch and range systems.

Training units are those that perform training activities to attain proficiency in the above space operations.

Tailored Questions to Assess Effects of Transferring Operational and Training Units to the Space Force on Organizational Attributes

To determine whether a transfer of these operational units and the training units that support them might enhance the effectiveness of the joint force, efficiency of the use of resources, independence of the Space Force, or sense of identity of the Space Force, we asked the following questions:

- Does the unit perform or directly support national, strategic missions, or joint operations?
- Does transfer of the unit to the Space Force help consolidate space operational or training activities in DoD to reduce duplication of efforts, leverage synergies, or enhance integration of space doctrine and capabilities?
- Does transfer of the unit to the Space Force help advance space warfighting doctrine, culture, or independent military options?

We then asked the following questions to assess whether a transfer from the current service might degrade effectiveness or efficiency:[1]

- Does the unit support missions that are not assigned to the Space Force?
- Does the unit perform or support activities that require service-unique training or expertise?
- Does the unit share infrastructure with other units that support nonspace missions?
- Does the unit operate in austere locations that require support that the Space Force might not be able to provide?

Results of RAND's Assessment of Space Operational Organizations

Table 4.1 provides a summary of our assessment of units and organizations that perform or directly support space operations.

The majority of space operational units in DoD were in Air Force Space Command (AFSPC). According to our analysis, all of AFSPC operational units should move, with some caveats. There are a few units in other Air Force organizations that should transfer, with the Air Force space intelligence organizations being the next-largest organization in the Air Force that should transfer to the Space Force. In the Army, most of the space operational units are assigned to U.S. Army Space and Missile Defense Command, Army Forces Strategic Command (USASMDC/ARSTRAT). Our analysis indicates that space operational units that support national, strategic, or joint missions should transfer, and those Army units or personnel whose primary mission is to provide space support to Army operations should remain in the Army. These would include some of the Army elements in NRO. The Navy's space operational units are primarily associated with operations of the narrowband SATCOM systems; thus, they should transfer to the Space Force, with some caveats. According to our analysis,

---

[1]   We assume that a transfer of an operational or training unit to the Space Force would likely not degrade the independence or the identity of the Space Force, but not transferring could degrade either of those attributes.

Table 4.1
Assessment Results for Transfer of Operational Activities and Units

	Activities and Units to Transfer to the Space Force	Activities and Units to Remain in Current Service or Organization
AFSPC	• 614 AOC, 21 OG, 50 OG, 50 NOG, 460 OG, 45 OG, and 30 OG (with agreement for Space Force to provide support to Air Force) • Air Force elements at NRO performing or directly supporting satellite operations or intel duties for the space domain	• Air Force elements at NRO supporting Air Force operations
Other Air Force	• 557th Weather Wing, 2nd Weather Squadron • Air Force Technical Applications Center, Detachment 26 • AFRL unit operating Maui Space Surveillance Site • Space-focused intelligence personnel in 544 ISRG and 659 ISRG • Space personnel in NASIC (with agreement for Space Force to leverage Air Force infrastructure)	• Not applicable
USASMDC/ ARSTRAT	• 1st Space Company (JTAGS) • 53rd Signal Battalion (WSOC) • RSSC (Space Force oversight, jointly manned) • Space control forces in 1st Space Brigade (for space control operations creating global or theater effects)	• 2nd Space Company (Army Space Support Team) • Army elements at NRO supporting Army operations • Space control forces in 1st Space Brigade (for space control operations creating local effects) • USASMDC/Reagan Test Site
Navy	• Naval Satellite Operations Center • MUOS network management • Naval Research Lab Blossom Point Tracking Facility	• MUOS teleports (Space Force oversight plus agreement for Space Force to leverage Navy personnel) • Space control forces creating local effects • Navy elements at NRO supporting Navy operations
Other DoD	• MDA satellite TT&C operations	• MDA satellite mission planning

NOTE: AOC = Air and Space Operations Center; OG = operations group; NOG = network operations group; AFRL = Air Force Research Laboratory; ISRG = Intelligence, Surveillance, and Reconnaissance Group; NASIC = National Air and Space Intelligence Center; JTAGS = Joint Tactical Ground Station; WSOC = Wideband Satellite Communications Operation Center; RSSC = Regional Satellite Communications Support Center; MUOS = Mobile User Objective System; MDA = Missile Defense Agency; TT&C = telemetry, tracking, and command.

Navy personnel or units that provide space support to Navy operations should remain in the Navy, including the Navy elements in NRO. Any remaining satellite operations in the Navy and other DoD organizations should also transfer and consolidate within the Space Force. We discuss more-detailed justifications and caveats in subsequent sections.

Air Force Space Command

Our analysis indicates that the operational units in AFSPC should transfer to the Space Force, with one caveat. The 614th AOC and operations groups in 21st Space Wing, 30th Space Wing, 45th Space Wing, 50th Space Wing, and 460th Space Wing should transfer because they solely perform or directly support space missions assigned to the Space Force, with the exception of the 2nd Range Operations Squadron in the 30th Operations Group. The 2nd Range Operations Squadron conducts range operations to support intercontinental ballistic missile (ICBM) tests for Air Force Global Strike Command (AFGSC) in addition to supporting space launch operations.[2] We judge that it is important for the Space Force to control the end-to-end process associated with the space launch operations; thus, the 2nd Range Operations Squadron should transfer because range operations are a critical support activity to launching space systems. Although ICBM tests support critical strategic missions for the Air Force, AFGSC is currently a customer of this service from AFSPC; thus, we judge that a similar relationship can be established between the Space Force and the Air Force by establishing a formal agreement, such as a memorandum of agreement (MOA) to leverage this service from the Space Force.

According to our analysis, the Air Force elements at NRO that are performing or directly supporting satellite operations or intelligence duties to support space operations should transfer to the Space Force. Such a transfer would help consolidate space expertise in DoD and could enhance the training of space personnel. This transfer implies that the current agreement and relationship between the Air Force and NRO would transfer to the Space Force, and those personnel would continue to support NRO as Space Force personnel. The Air Force should consider retaining some of these Air Force elements at NRO to support its own operations and needs.

Other Air Force

We identified several units in the Air Force outside AFSPC that perform or directly support space missions assigned to the Space Force (see Table 4.1). All these units should transfer to the Space Force, based on our assessment. Operations of the Maui Space Surveillance Site solely support the SSA mission.[3] Transfer of this unit would contribute to consolidating SSA activities in the Space Force. Consolidation could enhance training of SSA system operators and make the Space Force the center of excellence for SSA, which in turn could enhance the sense of identity of the Space Force. Air Force Technical Applications Center, Detachment 26, operates the nuclear

---

[2]   Ian Dudley, "Highway to Space," Air Force Space Command, January 15, 2016.

[3]   Air Command and Staff College, *AU-18, Space Primer*, Maxwell Air Force Base, Ala.: Air University Press, September 2009, p. 253; AFRL, "Air Force Research Laboratory Directed Energy Directorate Air Force Maui Optical and Supercomputing (AMOS) Site," fact sheet, 2017.

detonation payload on the GPS constellation.[4] This unit solely supports this national, strategic mission.

The 2nd Weather Squadron of the Air Combat Command's (ACC's) 557th Weather Wing performs space weather operations that support the SSA mission (as well as other military and civil missions) and space-based terrestrial weather operations that support all military forces.[5] Our analysis suggests that this unit should transfer to the Space Force because space weather information is a critical element of SSA, which enables space control operations—a vital mission for the Space Force in the establishment of its sense of identity and advancement of space warfighting doctrine. One area of concern is the relatively small size of the weather career field. Further analysis on the sustainability of the weather career field is needed to inform the transfer decision and the appropriate arrangements between the Space Force and the Air Force.

Intelligence personnel in the Air Force who conduct intelligence activities in support of space operations should transfer to the Space Force to enable close integration with space operations. Some of them are embedded in space operational units to support the operators; thus, when a space operational unit transfers to the Space Force, the intelligence personnel in that unit should also transfer. Separate intelligence units that support space operations are in 544 ISRG and 659 ISRG in ACC and in NASIC. Our analysis suggests that transferring these units to the Space Force would create a community that is focused on space and cultivate unique expertise in space intelligence, which could enhance effectiveness of space warfighting and enhance a sense of the Space Force identity. The Space Force, however, should leverage existing Air Force infrastructure at NASIC that is not unique to space (e.g., communications and computer networks) with an appropriate formal agreement in place, such as an MOA, thereby enhancing efficiency of the use of resources without degrading effectiveness.

Army

The majority of the Army's operational space units is in the 1st Space Brigade.[6] According to our analysis, Army units that perform space operations (as defined earlier) in support of joint operations should transfer to the Space Force, with some caveats, as we will discuss shortly. The 2nd Space Company (the Army space support teams) should remain in the Army because its mission is to provide space expertise to integrate space effects and employ them for Army operations.[7] Similarly, Army elements at NRO that exclusively support Army operations should remain in the Army. These units would require Army-unique expertise and training related to ground operations that the Space

---

4   Schriever Air Force Base, "Air Force Technical Applications Center, Det. 46," fact sheet, July 2016.

5   557th Weather Wing, "2nd Weather Squadron," webpage, undated.

6   Headquarters, Department of the Army, *Army Space Operations*, Washington, D.C., Field Manual 3-14, Change 2, February 13, 2018.

7   Headquarters, Department of the Army, 2018.

Force might not be able to provide. The 1st Space Battalion's 1st Space Company operates JTAGS to provide missile warning to joint forces in theater. It performs this mission using data from the same satellites (specifically, the Space Based Infrared System [SBIRS]) and applying the same analytical methods used to support the strategic missile warning mission.[8] We judge that transferring these units would unify the theater missile warning missions under the Space Force, which could enhance effectiveness, efficiency, and a sense of identity. It could foster cross-fertilization among the combined community of operators (JTAGS operators and SBIRS Mission Control Station operators), leverage commonality in training, enhance their training, and make the Space Force the center of excellence for them, fostering a sense of identity.

One caveat is that the supportability requirement for JTAGS needs to be considered. JTAGS is currently organized, trained, and equipped to operate in austere combat environments and may have unique supportability requirements to operate in combat zones, such as combat training; provision, setup, and maintenance of support equipment; and setup and maintenance of mission equipment. That said, JTAGS does not have to be a relocatable system operating in austere combat environments to be effective. It could be located on any main operating base within the theater, as long as communications links are established to disseminate its warning messages to users. If it is desirable to keep JTAGS as a relocatable system, the Space Force could seek alternative means, such as outsourcing to contractors or establishing an MOA with the Army, to receive support if it is not cost-effective for the Space Force to provide the support organically.

The 53rd Signal Battalion operates DoD wideband SATCOM payloads at the WSOCs to provide satellite transmission for the U.S. President, the Secretary of Defense, national agencies, DoD components, and joint commands.[9] Given the similarities in functions for operating and maintaining SATCOM satellites regardless of their operating frequencies (extremely high frequency, super high frequency, or ultra high frequency), consolidating satellite operation activities in the Army, Air Force, and Navy under the Space Force would unify the SATCOM missions, which could enhance the training of operators by leveraging commonality, and cultivate expertise in satellite operations.

Elements in the 4th Space Company and the 1st Space Brigade headquarters perform and support space control operations.[10] However, the Army's organization and CONOP for space control operations might evolve in the near term as the Army implements its multidomain CONOP. Thus, we refer to space control forces in the Army generically.

---

[8]   Dave Price, "JTAGS: Army Theater Missile Warning," *Air Defense Artillery*, October–December 2005.

[9]   U.S. Army Space and Missile Defense Command, "1st Space Brigade," webpage, undated.

[10] Headquarters, Department of the Army, 2018.

Our analysis indicates that the Space Force should have the responsibility for space control operations that create theater-wide effects, which affects joint operations in theater. Space control is a critical mission for the Space Force, as it is at the crux of space warfighting. Consolidation of space control forces could enhance cross-fertilization among a larger community of operators, which could in turn enhance the development of a common vision of the future of space warfare. Consolidation could further enhance the training of operators by leveraging commonality in training and equipment and could also make the Space Force the center of excellence for them, cultivating a sense that the Space Force members possess unique space experience and expertise.

However, the Army should retain some space control forces for conducting space control operations that create local effects, because space control forces may need to be in close proximity to the Army forces they are protecting or supporting to create an effect (e.g., counter-ISR). Such Army space control forces could have supportability requirements to operate in combat zones, such as combat training; provision, setup, and maintenance of support equipment; and setup and maintenance of mission equipment.

The other relevant Army space elements are the RSSCs and the Reagan Test Site (RTS). The Army operates and maintains the RSSCs, whose personnel conduct SATCOM resource planning (e.g., granting of access to SATCOM satellites and allocation of bandwidth) for combatant commands, services, agencies, and other users.[11] Personnel from the Air Force, the Navy, and the Defense Information Systems Agency also support the RSSCs. Transferring the RSSCs to the Space Force could enhance integration of satellite operations and planning activities and could enable the Space Force to implement an enterprise-wide approach to developing independent space warfighting doctrine and concepts relevant to SATCOM.

For instance, SATCOM planning and allocation processes at RSSCs need to be responsive and resilient to be effective when space is contested. The RSSC personnel would need appropriate training to operate in this new challenging environment, and the Space Force would likely have the relevant broad expertise to provide it. That said, according to our assessment of the impact on joint effectiveness, the oversight of RSSCs should transfer to the Space Force, but the RSSCs should remain as jointly manned centers because service-unique expertise may be needed (e.g., technical knowledge about specific SATCOM terminals for various airborne, ground, or maritime platforms or knowledge about user- or mission-specific SATCOM needs).

RTS is a critical test range supporting testing of missile defense systems, strategic ballistic missiles, and hypersonic systems for all services and MDA.[12] It also supports

---

[11] U.S. Army Space and Missile Defense Command and Army Forces Strategic Command, "RSSC: Regional Satellite Communications Support Centers," fact sheet, undated.

[12] U.S. Army Space and Missile Defense Command, "Ronald Reagan Ballistic Missile Defense Test Site at Kwajalein Atoll (RTS)," fact sheet, 2019.

two of the space missions assigned to the Space Force: space launch operations and SSA. There are limited space launch activities on Kwajalein, and thus we judge that it is likely to be more efficient for the Space Force to leverage the Army for this support and retain oversight of space launch operations. However, RTS collects critical SSA information because of its geographic advantage and suite of sensors. Inclusion of RTS in the Space Force would unify the SSA mission and could enable an integrated approach to SSA capability development. Further, although missile testing is RTS's primary mission, its sensors are likely to be used to support the SSA mission more routinely than the test mission because of the limited number of missile tests per year.[13] That said, RTS is one of many sensors in the Space Surveillance Network that support the SSA mission and thus it may not be as critical to the SSA mission as it is to the testing mission (especially when Space Fence comes on line in Kwajalein). On the basis of this assessment, RTS should remain in the Army and support the Space Force as a contributing Space Surveillance Network sensor.

Navy

The Navy's space operational units are primarily associated with operations of narrowband SATCOM satellites and their teleports and networks.[14] Our analysis suggests that operations and maintenance of the Navy's Satellite Operations Center and MUOS ground segment (MUOS teleports and the network management center) should also transfer to unify all SATCOM missions that support national, strategic, and joint force, with a caveat.[15] If the Navy unit that is supporting the MUOS teleport operations is sufficiently large and is dedicated to the MUOS teleport only, then this unit should transfer to the Space Force. However, Navy personnel performing the operations and maintenance of MUOS teleports may be supporting a variety of other communications-related activities on the naval station where the teleport is located. Thus, transferring operations and maintenance of MUOS teleports could reduce synergies with operations and maintenance of other ground antennas and communications networks at the same location. In such a case, oversight responsibility could transfer to the Space Force, with a formal agreement that allows the Space Force to leverage Navy personnel until a follow-on narrowband system replaces MUOS.

The Navy has another satellite operations center, Blossom Point Tracking Facility, which is managed by the Naval Research Laboratory (NRL).[16] It primarily focuses on TT&C of research and development satellites to support the Navy, other DoD users,

---

[13] MDA, "Integrated Flight Tests at U.S. Army Kwajalein Atoll/Ronald Reagan Ballistic Missile Defense Test Site (USAKA/RTS) Environmental Assessment," August 12, 2012.

[14] Adam DeJesus, "Navy Space Cadre and Warfighters Putting Pieces Together," *CHIPS*, June 10, 2014.

[15] We did not review operations of other narrowband SATCOM satellites that the Navy operates in detail, because they are past their original design life and are likely to be decommissioned in the near term.

[16] NRL, "Blossom Point Tracking Facility (BPTF)," fact sheet, April 2019.

and government agencies. Transferring this facility and personnel to the Space Force would help consolidate DoD satellite operation activities, which could increase effectiveness and efficiency by leveraging commonality in training and equipment.

Our research did not include an in-depth assessment of the Navy's space control activities. As discussed earlier, our analysis indicates that the oversight and conduct of space control operations creating theater-wide effects should be consolidated in the Space Force. Thus, space control operations for creating local effects to protect organic Navy units should remain as the Navy's responsibility. Additionally, Navy elements at NRO that exclusively support Navy operations and needs should remain in the Navy.

Other DoD

MDA operates the Space Tracking and Surveillance System, which consists of two satellites. It is an experimental system to support development of future missile defense space trackers.[17] MDA is also considering a space sensor layer in the future for tracking long-range ballistic and hypersonic missiles.[18] According to our analysis, the TT&C functions of the Space Tracking and Surveillance System (and any satellites that MDA might manage in the future) should transfer to the Space Force to help consolidate satellite operation activities and leverage commonalities across various DoD satellite operation centers. Certain functions, such as mission planning for the satellites, may require MDA's expertise on missile defense; thus, such activities should reside under the purview of MDA. The Space Force and MDA would need an appropriate MOA or liaisons to coordinate their respective satellite operation activities. That said, the Space Force should expect that establishing such a formal arrangement and relationship could be difficult and take time.

Results of RAND's Assessment of Training Organizations

Table 4.2 summarizes the results of our assessment of space training units or organizations. The majority of space training in DoD is provided by the Air Force and USASMDC. In general, training units that are relevant to those operational units should transfer to the Space Force to enable close integration with the operations. We discuss more-detailed justifications and caveats in subsequent sections.

Air Force Space Command

AFSPC's operations groups (21 OG, 30 OG, 45 OG, 50 OG, and 460 OG) have operations support squadrons that provide unit training for space operators, and our analysis suggests that they should transfer to the Space Force. The 319th Combat Training Squadron provides advanced individual training for space operators and should trans-

---

[17] MDA, "Space Tracking and Surveillance System," fact sheet, 17-MDA-9108, March 27, 2017.

[18] Jason Sherman, "MDA Identifies Nine Companies Competing for Space Sensor Layer," *Inside Defense*, March 7, 2019.

Table 4.2
Assessment Results for Transfer of Training Activities and Units

	Activities and Units to Transfer to the Space Force	Activities and Units to Remain in Current Service or Organization
AFSPC	• 319th Combat Training Squadron • 614th Combat Training Squadron	• Not applicable
Other Air Force	• 533rd Training Squadron (agreement with AETC to leverage facilities) • 25th Space Range Squadron • 527th Space Aggressor Squadron (agreement to support Air Force and joint training) • 328th Weapons Squadron[a] • National Security Space Institute	• 381st Training Support Squadron (agreement with AETC)
USASMDC and ARSTRAT	• Army Space and Missile Defense School (space portion)[b]	• Army Space and Missile Defense School (training on employment of space into Army operations)

NOTES: AETC = Air Education and Training Command. We did not identify relevant training units or organizations in the Navy, because Navy space operators meet their training requirements through on-the-job training and space-related courses offered outside the Navy. The Naval Postgraduate School is excluded from our analysis because we did not address organizations that provide individual space education.

[a] Should consider a direct reporting unit to Headquarters, Space Force.

[b] The schoolhouse also provides training courses for missile defense operations.

fer to the Space Force.[19] The 614th Combat Training Squadron is primarily providing training and certification to work in the 614th AOC and should accompany the transfer of the 614th AOC to the Space Force.[20]

Other Air Force

Several units in AETC and in ACC provide or support space training, and, according to our analysis, they should transfer to the Space Force to consolidate space training activities, cultivate space expertise, enhance a sense of identity for the Space Force members, and control the training of Space Force personnel.

AETC's 533rd Training Squadron, which provides initial qualification training for satellite operations, should transfer to the Space Force, but the Space Force should

---

[19] The 319th Combat Training Squadron was formerly the Advanced Space Operations School. The 319th Combat Training Squadron expands space system understanding by providing instruction of space systems, capabilities, requirements, acquisition, strategies, and policies to support joint military operations and U.S. national security.

[20] The 614th Combat Training Squadron is organized into seven flights: training, standardization and evaluations, weapons and tactics, exercises and experiments, systems integration, space weather, and special technical operations support. See Nicholas Mercurio, "New Squadron, 614th CTS, Activates at Vandenberg," Vandenberg Air Force Base, Calif., December 21, 2016.

leverage the facilities and generic services that AETC provides.[21] The 381st Training Support Squadron offers basic instructor courses for faculty training and other services not unique to space, such as registrar services and facilities management. According to our analysis, this unit should remain in the Air Force, and the Space Force should leverage these services to train its instructors to minimize duplication and enable efficient use of resources.

The 527th Space Aggressor Squadron is part of the 57th Adversary Tactic Group, and it supports military training exercises involving air and joint operations in contested space environments by conducting adversarial tactics.[22] To cultivate expertise in space control operations in the Space Force and to foster development of space warfighting concepts and doctrine, this unit should transfer. The Space Force should establish formal agreements to continue provisioning space aggressor support to the Air Force and other services.

The Air Force Weapons School's 328th Weapons Squadron provides advanced training for space operators to improve their tactical knowledge (e.g., defensive space control tactics) and supports the development of advanced tactics and tactical doctrine.[23] The squadron would play an important role in advancing space warfighting doctrine and concepts and thus should become a direct reporting unit to Headquarters, Space Force.

The 25th Space Range Squadron is part of the Nevada Test and Training Range, and it operates a space test and training range for DoD (e.g., planning and conducting a joint exercise involving space control operations).[24] This unit should transfer to the Space Force to cultivate space expertise and to enable close integration with space training and test activities.

The National Security Space Institute (NSSI) is under the Air Force Institute of Technology, and its primary mission is to provide continuing space education to space professionals in DoD.[25] However, it does offer fundamental space courses that are part of space training requirements for space personnel in the Air Force and the Navy. Our analysis suggests that NSSI should transfer to the Space Force, but the Space Force should leverage the Air Force Institute of Technology for administrative functions and other generic services that are not unique to space (e.g., communications and computer networks) for operations of NSSI.

---

[21] Vandenberg Air Force Base, "381st Training Group," fact sheet, December 6, 2016.

[22] The 527th SAS has three mission sets: GPS electronic attack, SATCOM electronic attack, and orbital engagement systems (Arielle Vasquez, "527th SAS: Preparing Warfighters Now, into the Future," Air Force Space Command, February 7, 2018).

[23] Nellis Air Force Base, "United States Air Force Weapons School," fact sheet, May 10, 2016.

[24] Schriever Air Force Base, "25th Space Range Squadron," fact sheet, June 22, 2018.

[25] NSSI, "About," webpage, undated-a.

Army

The main schoolhouse for Army space training is the Space and Missile Defense School.[26] According to our analysis, the Army space training activities for operations of JTAGS, SATCOM, and space control capability should transfer and consolidate with space training activities that transfer from the Air Force to the Space Force. Fundamental space courses that the Army space support teams and Army space control operators (i.e., space control operators exclusively supporting Army operations) should also be consolidated with the courses to be offered in the Space Force to minimize duplication of efforts. Further, the Space Force is likely to have the most up-to-date information relevant to space operations (e.g., space threats, space capabilities, and space tactics). The Army space professionals should receive space-relevant training from the Space Force, and an Army liaison could provide Army perspectives to the space curricula.

Navy

The Navy does not have dedicated space training units or organizations for its space operators. In the Navy, space operations is not a separate career field (or designator) but a subspecialty (or additional qualification designator). Navy space operators meet their training requirements through on-the job training in space-related billets and through courses offered at NSSI and the 319th Combat Training Squadron.[27] The Navy liaisons to NSSI and the 319th Combat Training Squadron serve as instructors, providing Navy perspectives and Navy-relevant information.[28] According to our assessment, the Navy should continue to maintain this liaison role once NSSI and the 319th Combat Training Squadron transfer to the Space Force. The Naval Postgraduate School offers space-related courses for individual space education, but our analysis did not address education organizations.

---

[26] The Space and Missile Defense School's space institutional training courses (some of which may be future courses) include the Space Operations Officer Qualification Course, Space Senior Leader Seminar, Army Space Cadre Basic Course: Satellite Communication Electromagnetic Interference Fundamentals Course, Electromagnetic Interference Characterization Course, Army Space Control Fundamentals Course, Mobile Integrated Ground Suite Initial Qualification Training, Advanced MIGS Operations, Advanced Space Control Systems Courses, Army Space Control Command and Control Course, Space Control Planners Course, Tactical Space Operations Courses, Joint Tactical Ground Station IQT, and JTAGS Leader Development Course. See Dottie K. White, "U.S. Army Space and Missile Defense School House Earns 'Institute of Excellence' from TRADOC," U.S. Army, November 15, 2016.

[27] Matthew E. Faulkenberry, "Critical Review of the Navy Space Cadre," thesis, Monterey, Calif.: Naval Postgraduate School, June 2014, p. 34.

[28] NSSI, "Partners," webpage, undated-b.

Acquisition Activities

As mentioned in Chapter Three, the acquisition authority for Space Force systems will remain with the secretariat in DAF. The acquisition-related activities we assess in this section are those that the Air Force, other services, and DoD organizations currently perform and what the Space Force might perform in the future.

We organized our assessment of what acquisition-related activities should transfer to the Space Force to match the approach of the SFPTF Acquisition Planning Team,[29] using the following work breakdown structure:

- **Oversight and advocacy:** command and control responsibilities in Headquarters, Space Force, and the headquarters responsibilities to communicate about the Space Force's mission to external parties, particularly Congress, the White House, the Office of the Secretary of Defense, the Joint Staff, and the general public
- **Requirements and architecture:** the Space Force's (1) development of capability requirements as a participant in Joint Capabilities Integration and Development System (JCIDS) processes and (2) translation of requirements into a portfolio of system development, production, and sustainment activities
- **Policies and procedures:** development of warfighting doctrine and its translation into policies and procedures that Headquarters, Space Force, could use to command and control activities throughout the service
- **Research, technology development, and innovation:** investment in science and technology that can form the basis for creating new programs of record or fielding urgently required new capabilities
- **Programs of record:** management of the materiel and information system programs that develop, produce, and sustain the systems that operational units in the Space Force would employ
- **Test and evaluation:** execution of the development and operational test and evaluation for programs of record
- **Sustainment:** execution of sustainment of programs of record
- **Rapid acquisition:** special arrangements through the offices of the secretary of the Air Force or the under secretary for the Space Force that facilitate rapid acquisition of high-priority system acquisition programs; by reducing the number of layers between a program and the senior leadership and by promoting close, crossfunctional coordination within any layer, these arrangements cut the calendar time and the level of resources required to approve decisions about acquisition

---

[29] Our work breakdown structure for space acquisition is a slight variation on what is reported in Robert Clausen, Chris Elenbaum, Tim Sejba, and Brendan Hochstein, "Organizing U.S. Space Force Acquisition," briefing, Washington, D.C.: Space Force Planning Task Force Acquisition Planning Team, Headquarters, Air Force, May 9, 2019.

- **Career field management:** management of all personnel in the defense acquisition workforce who play roles in the activities above.[30]

Tailored Questions to Assess the Effects of Transferring Acquisition-Related Authorities and Responsibilities to the Space Force on Organizational Attributes

For each of the acquisition-related activity areas described above, we asked the following questions to assess effects on the four organizational attributes:

- How does current formal guidance assign authority and responsibility?
- How important is an acquisition activity to the effectiveness of Space Force missions?
- Does an acquisition-related activity display economies of scale relevant to effectiveness or efficiency that cannot be realized fully within the Space Force?
- Can a high-level executive accelerate acquisition for high-priority systems by simplifying bureaucratic coordination?
- Given formal authority, what arrangements increase the independence and identity of the Space Force without imposing undue bureaucratic inefficiency?
- Can the workforces in acquisition-related activities sustain effective horizontal coordination and limit communication through vertical stovepipes to affect decisions?

Results of RAND's Assessment of Acquisition Organizations

To be clear, acquisition authority remains with the secretariat, but acquisition-related organization in the Space Force will be staffed with Space Force personnel, and any authorities or command relationships that had reported to the Air Force or another service will instead be routed to the Space Force. We often suggest that activities within a large organization should transfer but do not attempt to identify which specific parts of that large organization might accompany that transfer. Table 4.3 provides an overview of our results, which are explained in more detail in the remainder of this section.

Air Force Space Command

According to our analysis, all acquisition-related activities in AFSPC should transfer. These include, for example, Headquarters, AFSPC; the Space and Missile Systems Center; the DoD portion of the Space Security and Defense Program; and the Space Rapid Capabilities Office. The Space Force's warfighting capabilities will depend directly on the material and information systems that it acquires, and these organizations will directly affect decisions about what to buy and how to buy it. The Space Force's control of these organizations could be critical to its independence. The closer their acquisition is to the organizations that would actually employ the systems in a

---

[30] We delay discussion of this element of acquisition until the discussion of workforce management (in Chapter Five).

Table 4.3
Assessment Results for Transfer of Acquisition Activities and Organizations

	Activities and Organizations to Transfer to the Space Force	Activities and Organizations to Remain in Current Service or Organization
AFSPC	• All acquisition-related activities (e.g., Headquarters, AFSPC; Space and Missile Systems Center; Space Security and Defense Program; and Space Rapid Capabilities Office) • Air Force elements at NRO conducting acquisition activities	• Not applicable
Other Air Force	• Air Force Operational Test and Evaluation Center Detachment at Peterson Air Force Base	• AFRL (Funds and billets from space directorates in AFRL transfer to Space Force; Space Force applies the funds to use Air Force personnel and infrastructure to perform work) • Test and evaluation on systems or components not specific to space (Air Force personnel and infrastructure to perform work for Space Force as needed) • DAF-wide Rapid Capabilities Office (for both Air Force and Space Force high-priority work)
USASMDC/ ARSTRAT	• USASMDC Future Warfare Center and Technical Center's roles specific to space move to Space Force with resources • ARL's roles specific to space move to Space Force and AFRL with resources	• USASMDC (acquisition activities not related to space and acquisition personnel advocating for Army space requirements)
Navy	• The roles of Space and Naval Warfare Systems Command (now NIWC) specific to space move to Space Force with resources, including MUOS program of record • NRL's roles specific to space move to Space Force and AFRL with resources	• NIWC (acquisition activities not related to space, acquisition personnel advocating Navy space requirements)
Other DoD	• SDA • MDA (space sensor layer acquisition activities); good arguments also exist for remaining in MDA	• DARPA and SCO (limited space work remains where it is—relationships relevant to space-wide initiatives move from Air Force, Army, and Navy to Space Force) • MDA (acquisition activities not related to space)

NOTE: ARL = Army Research Laboratory; NIWC = Naval Information Warfare Center; SDA = Space Development Agency; DARPA = Defense Advanced Research Projects Agency; SCO = Strategic Capabilities Office.

contested space environment, the more likely these systems are to contribute effectively to joint capabilities.

Current Air Force elements at NRO conducting acquisition activities should transfer to ensure effective and efficient coordination with their counterparts within the Space Force. Space Force control of these elements is also likely to ensure an independent Space Force voice in multiparty discussions and negotiations within NRO.

Other Air Force

Air Force Operational Test and Evaluation Center Detachment 4, at Peterson Air Force Base, is located next to space operators, allowing testing in an operationally realistic environment. Our analysis suggests that transfer of this detachment is likely to ensure that the priorities of operators in the Space Force inform decisions about capabilities most relevant to fielding. The current presence of Detachment 4 with space-oriented activities at Peterson also suggests that this is likely to be an effective and efficient place to locate these test capabilities.[31]

Space-related research at AFRL should remain within the Air Force to take advantage of the current infrastructure and to provide personnel who are conducting space-related research access to a broader intellectual environment. Defense research laboratories sustain flexibility by making it easier to match available organic personnel to demands for research as those demands change over time. Because such matching is easier and more effective in larger laboratories, collocating researchers with space-related skills with Air Force research activities should improve the effectiveness and efficiency of their use. Too much distance from operators, of course, could make researchers with space-related skills less responsive to operators' priorities. The Space Force should be given a role in the governance of AFRL to ensure that its priorities would be reflected in the research that AFRL supported. To strengthen the governance role, the Space Force should receive funding that it could apply to specific activities within AFRL. Thus, the funding and billets associated with AFRL's space directorates should transfer to the Space Force for it to control, but the Space Force should apply this funding to use AFRL personnel to execute research activities that the Space Force chooses each year.

In this arrangement, personnel working on space-related research would reside in AFRL under the Air Force and be managed over their careers by the Air Force. Although the Space Force would not own the personnel it uses, it would generate program objective memoranda for the funds used to pay them and closely control the work that they do for the Space Force.[32] AFRL and the Space Force should negotiate which AFRL researchers will work on Space Force topics each year.[33] Thus, the person-

---

[31] Cameron S. Hunt, "AFOTEC Maintains Mission Readiness Through Operational Testing," U.S. Air Force, February 26, 2019.

[32] We do not address how the Space Force might pay for nonlabor costs within AFRL.

[33] If the Space Force is not satisfied that AFRL can provide organic personnel with the right skills, experience levels, and performance records, the Space Force should reserve the right to substitute contractor personnel with the right qualifications to work at AFRL or elsewhere, as appropriate.

nel working for the Space Force could change from year to year as the Space Force's needs change, even if they spend their whole careers working at AFRL.

This arrangement would allow the Space Force to have strong control over space-related research and development (R&D) work at AFRL and provide the flexibility to access the right people with appropriate skills from throughout AFRL.[34] Further, it gives the Space Force greater flexibility to adjust its research portfolio from year to year without interrupting the career management of the people it wants to access.[35] This arrangement is much easier to implement for government civilian researchers than for military researchers. That said, our expectation is that the AFRL researchers the Space Force would access would mainly be civilians.

Test and evaluation of systems or components of systems that are not unique to space should remain in the Air Force to take advantage of scale economies there. Uncertainty about the conduct of testing, particularly development testing, creates variation in demands on test facilities that can best be accommodated in larger test facilities that can afford to maintain more spare capacity to absorb such variation. This capacity includes physical infrastructure and the personnel with technical skills unique to testing.

Even if the Space Force maintained its own rapid acquisition capabilities, it should retain its access to the Air Force Rapid Capabilities Office (AFRCO). AFRCO, by definition, allows a high-level focus on only a small number of initiatives. But if an initiative relevant to the Space Force has a high enough priority within DAF to warrant the secretary's direct attention, AFRCO should continue to give programs within the Space Force the secretary's attention. For high-priority programs that do not warrant such attention, the Space Force could maintain its own rapid capability office.

Army

The Army's USASMDC currently addresses some space-related issues in its Future Warfare Center and Technical Center; this work accounts for only a small share of the total activity at these centers. According to our assessment, these centers should remain in the Army, but work specific to space should transfer to the Space Force. Such a transfer would create new seams between space-based capabilities and land-based activities that these capabilities would support, potentially hurting the effective-

---

[34] For a useful discussion of many of these issues, see Drake Warren, Heather Krull, Jennifer Lamping Lewis, Aisha Najera Chesler, Ellen M. Pint, and J. Michael Gilmore, *Assessment of Alternative Funding Models for Activities in RDECOM (Now CCDC) and ATEC*, Santa Monica, Calif.: RAND Corporation, RR-2818-A, 2020.

[35] We assessed that the value of (1) giving researchers access to a broader intellectual environment and (2) the flexibility provided by leaving AFRL space directorates in the Air Force outweighs any benefits to the Space Force of owning the personnel who conduct space R&D. However, decisionmakers might not necessarily make the same trade-offs. Should independence in space R&D activities be critically important to the decisionmakers, then the Space Force should directly manage and control the funding and personnel associated with AFRL's space directorates, recognizing that there is a risk that the Space Force research lab might not be adequately flexible to respond to surprises.

ness and efficiency of the joint force. But a key presumption behind the creation of the Space Force is that the integration of space capabilities across DoD would eliminate enough seams to offset the effects of the new seams created between the Space Force and the rest of DoD.

We are not aware of empirical evidence to support this claim; we accept it as inherent in the guidance to create a Space Force that defines the scope of our analysis. If evidence revealed over time that moving this work to the Space Force does not yield the dividends currently claimed, this transfer decision should be revisited. In the meantime, the Space Force must develop its own warfighting doctrine and a portfolio of technologies to support it. USASMDC resources relevant to space are a reasonable place to look for resources to support these efforts.

Most work at USASMDC is focused on missile defense, not on space per se. That work should remain in USASMDC within the Army. And if the Space Force takes on new roles in space, the Army would need to commit resources to define its relationship with the Space Force and to articulate its priorities with regard to space capabilities in JCIDS processes. USASMDC is a natural place to do this in the Army. It should retain the capabilities it needs to do this.

ARL currently conducts a small amount of research on space. As the Space Force takes on the responsibility of integrating space-related capabilities, AFRL could provide the infrastructure and community of researchers to support the kind of research that the Army currently does on space. The funds and billets associated with this research would move from the Army to the Space Force, where they would be applied to space-related research at AFRL. It might be cost-effective to give USASMDC a role in the governance of this research in AFRL. USASMDC should also be able to contribute funding to support relevant research in AFRL and, in particular, to offer its assets at Kwajalein Atoll to support that research when appropriate. Conversely, the Space Force should have funding to pay the Army for work at Kwajalein Atoll when appropriate. New agreements among the Army, Air Force, and Space Force would be helpful.

Navy

NIWC (formerly the Space and Naval Warfare Systems Command) Systems Center conducts a small portion of its work on space capabilities. The Navy has traditionally viewed the portion of space above the oceans as an integral part of its warfighting domain, so it is natural that it would pursue capabilities in space. But, as in the case of the Army, if the Space Force takes on the responsibility of integrating capabilities in space, our analysis suggests that such Navy initiatives should move to the Space Force. This would create new seams between the Navy and Space Force that could degrade the effectiveness and efficiency of the joint force.

We take the guidance to create the Space Force—in part precisely to create such seams—as a presumption that the seams that the Space Force would eliminate within the space domain would more than compensate for the creation of these new seams. In addition, if the transfer of the Navy's space capabilities to the Space Force does not generate the gains projected for the joint force, the decision to transfer these capabilities should be revisited. The effects of this transfer would likely be largest in the transfer of the MUOS satellite communication program of record. It currently serves users across DoD, but Navy users benefit most from it.

NIWC should remain in the Navy, where it would continue to conduct the large majority of its work, which does not currently relate to space. As with the Army, if the Space Force takes on a new role in space, the Navy will need to commit resources to define its relationship with the Space Force and to articulate its priorities with regard to space capabilities in JCIDS processes. The Navy should retain the capabilities it needs to do this.

NRL currently conducts a small amount of research on space. If the Space Force takes on the responsibility of integrating space-related capabilities, AFRL can provide the infrastructure and community of researchers to support the kind of research that the Navy currently does on space. The funds and billets associated with this research would move from the Navy to the Space Force, where they would be applied to space-related research at AFRL. It might be cost-effective to give the Navy a role in the governance of this research in AFRL. As with the Army, new agreements among the Navy, Air Force, and Space Force would be helpful.

Other DoD

SDA, currently under the oversight of the under secretary of defense for research and engineering, is expected to transfer to the Space Force if the Space Force is created.[36] We endorse that transfer. The Space Force is being created to increase DoD's focus on space-related warfighting issues. SDA currently has a similar intent and could enforce the Space Force's ability to assert its independence as it develops new doctrine and capabilities.

Space-related research in DARPA should remain in DARPA. DARPA commits small amounts of resources to offer qualitatively new perspectives on DoD activities that have challenged traditional perspectives. Even if the Space Force takes on primary responsibility for warfighting in a contested space domain, space-related efforts in DARPA could provide a valuable perspective on space capabilities that could complement the new efforts of the Space Force. In its past efforts, DARPA has often partnered with the armed services and passed innovations to them. The Space Force should seek opportunities to do this in the future, even if challenges sometimes slow the transition of new ideas to the Space Force.

---

[36] Sandra Erwin, "Shanahan Officially Establishes the Space Development Agency," *Space News*, March 13, 2019.

If SCO migrates to DARPA as currently planned, the statements above would apply to it also.[37] SCO would presumably take on a more DARPA-like approach than the one it uses today. If it retains its current mission of fielding capabilities relevant to combatant commanders within two to five years, it is likely to benefit from remaining independent of the Space Force, even if it is fielding a system for U.S. Space Command, the combatant command of greatest importance to the Space Force. For the foreseeable future, the Space Force is likely to be heavily occupied with creating a longer-term identity that could sustain its responsibilities in space. Allowing SCO to take on high priorities that have very short-term decision horizons could help the Space Force stick to its longer-term mission. Agreements with SCO should help ensure that this occurs.

MDA is developing a space sensor layer that could fall under the purview of the Space Force if it integrates space capabilities across DoD. This is part of an MDA effort to develop a highly time-sensitive, integrated kill chain that would combine capabilities located in space and elsewhere. A new seam between the Space Force and assets that remain in MDA could potentially prevent effective implementation of this new kill chain. We believe that this MDA activity deserves closer analysis before a decision is made to transfer it to the Space Force or retain it within MDA.[38] Acquisition activities within MDA that are not related to space should remain in MDA.

Major Headquarters Activities

A number of headquarters activities can affect the Space Force's ability to communicate independently and advocate for its priorities in the broader setting of DoD. They include activities associated with the following:

- **Concepts and doctrine:** Current Air Force warfighting concepts and doctrine emphasize airpower and see space capabilities primarily in a supporting role. Advocates of a Space Force argue that Air Force concepts and doctrine have prevented capabilities in space from becoming any more than enablers of military capabilities to act outside space. An independent Space Force could, in principle, offer a broader view of space. But the Space Force would have to control its own warfighting concepts and doctrine for such a view to emerge.
- **Requirements:** JCIDS processes prioritize capability requirements and then assign responsibilities for developing them across DoD. Without appropriate requirements, the Space Force could not create an effective force structure or justify its need for resources to organize, train, and equip its force structure.

---

[37] Colin Clark, "Top DoD Official Shank Resigns; SCO Moving to DARPA," *Breaking Defense*, June 17, 2019.

[38] Other options are possible. For example, MDA could develop sensors and prototypes that, if development were successful, the Space Force could procure, operate, and sustain. Guidance for development should be clear to ensure that MDA took an appropriate life-cycle perspective that included goals for cost-effective sustainment.

The Space Force would likely have to control its participation in JCIDS processes to ensure that (1) DoD requirements reflect the capabilities relevant to the Space Force's understanding of warfighting in a contested space environment and (2) the Space Force has the authority to build the force structure it needs to pursue its view of such warfighting.

- **Planning, programming, budgeting, and execution system (PPBES):** Validated requirements open the door for the Space Force to compete for resources with other parts of DoD. If the Space Force relies on the Air Force to represent its priorities in the broader DoD resourcing process, the leadership of the Air Force could (in the eyes of advocates for the Space Force) once again fail to understand the relevance of the Space Force's view of warfighting to the joint capabilities of DoD as a whole. To realize the force that its warfighting concept and doctrine envision, the Space Force will likely need the authority to defend its views in the broader political process that is planning, programming, and budgeting in DoD.

- **Legislative liaison:** DoD can propose a budget with the participation of all its components, including (potentially) the Space Force. Congress ultimately finalizes each budget. The Space Force will likely need the authority to defend its portions of the DoD budget when the proposed budget went to Congress. The Space Force will ultimately need to compete in a congressional setting with any other organization within DoD that might be given the authority to represent the interests of the Space Force on Capitol Hill. The secretariat of the Air Force has the formal authority to maintain effective liaison activities between the Space Force and Congress. Arrangements should be made to preserve an independent voice for the Space Force within these liaison activities to ensure that more traditional views from the Air Force leadership do not dominate new perspectives from the Space Force.

- **Public affairs:** To the extent that Congress makes decisions designed to build public support, the Space Force could presumably influence Congress to communicate directly with the general public. Again, the Space Force will need to compete for the public's attention with any other part of DoD that might be assigned to represent the Space Force's views. It could communicate most independently with the general public by being able to communicate directly with the public, on its own behalf.

The more control that Headquarters, Space Force, could exert over these activities, the more effectively it could pursue its independent view of how to fight wars successfully in a contested space environment. However, current policy guidance discourages the transfer of several of these activities to the Space Force.

Tailored Questions to Assess Effects of Transferring Headquarters Activities to the Space Force on Organizational Attributes

For each of the headquarters activities described above, we asked the following questions about the potential effect of transferring it to the Space Force:

- Does current guidance assign authority or responsibility that degrades the independence or sense of identity of the Space Force?
- Would an independent Headquarters, Space Force, activity impose undue bureaucratic inefficiencies?
- Does a headquarters activity display economies of scale relevant to effectiveness or efficiency that could not be achieved within Headquarters, Space Force?

When the transfer of a headquarters activity to the Space Force challenged current guidance or degraded effectiveness or efficiency, we asked how Air Force and Space Force authorities might be tailored to ameliorate these effects.

Results of RAND's Assessment of Critical Headquarters Elements

The suggested transfers of headquarters activities that result from our analysis are expected to help enhance the independence and sense of identity of the Space Force. Because the Space Force is being advocated in part as a mechanism to create a new, independent voice in DoD, we assume that this organizational attribute deserves special emphasis when weighed against other attributes.

However, these transfers would likely reduce the efficiency of the administrative processes that DAF would use to manage these activities. This could occur because the suggested transfers create some duplication of effort in support of the Space Force and the rest of DAF. They also could limit direct horizontal communication between the Space Force and the rest of DAF and, as a result, force coordination to occur through vertical channels that would push decisions up through the Space Force and through the rest of DAF for resolution at the highest level within DAF as a whole.

But the administrative changes that would induce such inefficiency are likely to be small relative to the positive effects that they would likely have on the independence and sense of identity of the Space Force. The administrative processes involved would likely be very small in the broader scope of the Space Force and DAF as a whole. And the duplication and barriers to communication would likely have relatively small effects even relative to each administrative process affected.

Table 4.4 summarizes the results of our assessment of headquarters activities. Details on particular headquarters activities follow.

Concepts and Doctrine

According to our analysis, the CSO should have direct control over the development of warfighting concepts and doctrine. The secretariat of the Air Force could facilitate coordination between Space Force and Air Force concepts and doctrine, but the sec-

Table 4.4
Assessment Results for Transfer of Headquarters Activities

Headquarters Elements	Activities to Transfer to the Space Force	Activities That Do Not Transfer
Concepts and doctrine	• Under the direct control of the CSO	• Air Force concepts and doctrine; coordination with Space Force concepts and doctrine
Requirements	• Use concepts and doctrine to translate capability requirements into DOTMLPF requirements for the Space Force • Translate material requirements into engineering requirements for Space Force systems • Manage translation of engineering requirements into specifications to acquire systems	• Once joint processes assign a requirement to the Space Force, the Air Force has no further role in that requirement

PPBES	• Build program, and budget to allocate resources against a Space Force total obligation authority target
	• Advocate the Space Force program and budget in the Air Force PPBES processes and congressional review
	• Execute the Space Force budget within the Air Force financial management processes
		• Secretariat of the Air Force administers a PPBES process to
		  – service Space Force and Air Force programs
		  – protect the Space Force's independent ability to allocate resources within its budget and advocate for its own program and budget
		• Secretariat of the Air Force administers integrated financial management processes for the Air Force and Space Force

Legislative liaison and public affairs	• Control independent advocacy of Space Force issues within DAF legislative liaison and public affairs offices at all levels	• Secretariat of the Air Force leads legislative liaison and public affairs offices without preventing Space Force independence

NOTE: DOTMLPF = doctrine, organization, training, materiel, leadership and education, personnel, and facilities.

retariat would not likely have sufficient resources of its own to take on full integrative responsibilities. The Space Force and Air Force would need to take responsibility for coordinating their relationship in issues of mutual interest. Our goal here is to give the Space Force as much authority as possible to realize the potential associated with the decision to create it. The Air Force should defer to the Space Force for concepts and doctrine associated with space activities, giving the Space Force room to grow and mature on its own terms. The Space Force, of course, should ultimately be held accountable for the concepts and doctrine that this open approach allowed it to develop.

Requirements

Current policy guidance envisions the Space Force as an equal partner in JCIDS processes. This role would, in principle, give the Space Force full reign to translate its concepts and doctrine, as it develops them, into formal positions that it could advance in the Joint Requirements Oversight Council (JROC). When capability requirements are assigned by JROC, the Space Force could weigh the full range of options it might use to address new capability requirements. Those approved for material solutions could then form the basis for (1) investments in science and technology and (2) engineering specifications that the Space Force could then pursue in programs of record.

If we draw on the presumption that the Space Force has been proposed, in part, to identify opportunities in space that the Air Force has not identified in the past, we can assume that a Space Force should exercise these requirements-related responsibilities with as much independence from the Air Force as possible. Once the JROC assigns a requirement to the Space Force, the CSO would take and retain the lead on managing all activities that flowed from this assignment. The Space Force and Air Force would act as separate armed services as they executed the tasks required to integrate their respective requirements-related responsibilities.

Planning, Programming, Budgeting, and Execution System

Current formal guidance states that the secretariat of the Air Force will continue to administer all planning, programming, and budgeting activities for the Air Force and the Space Force. Scale economies associated with these processes clearly favor department-wide management over the creation of separate Space Force processes. However, to build ways to fight war in a contested space environment that the Air Force has not been able to create using the traditional approach to managing resources, the Space Force will need to control its force structure, the systems that it will fight with, and the way that it will use these systems—both in "peacetime" and when combatant commands call on the capabilities of the Space Force. But this will not likely be accomplished if the Space Force does not have effective control over its resources. To allow such control, the secretariat could give the Space Force effective control of key planning, programming, and budgeting *decisions* within the broader PPBES processes that the secretariat continues to *administer*.

In such an approach, the secretariat might identify a target for total obligation authority each year that the Space Force might then work within. The Space Force would build its program and budget within these annual constraints. Decisions about the Space Force would occur within secretariat programming and budgeting processes but would be segmented from other Air Force programming and budgeting so that competition for obligation authority applied only at the highest level within DAF. Ideally, the under secretary for space and chief of staff of the Space Force would oversee the Space Force portion of these activities and then submit final proposals for a program and budget to the secretary of the Air Force for final approval. When a final

DAF program and budget goes to the White House and then Congress for review and coordination, the Space Force should retain authority independent of the Air Force to advocate for its portions of these documents. The Air Force should continue to administer the structure of the programming and budgeting processes as it does today, but the process should be adjusted to ensure the segmentation of decisions between the Space Force and the rest of DAF, as described above.

Once the Air Force and the Space Force receive a budget from Congress, the Space Force should execute its portion of that budget within financial management processes that the secretariat will continue to maintain. Again, scale economies associated with financial management processes favor having the Air Force manage them over creating separate Space Force processes. But the Space Force should retain authority to use these administrative processes to move funds among its activities to the extent that Congress allows flexibility. Such flexibility should not extend to moving funds between activities in the Space Force and those elsewhere in the Air Force.

Legislative Liaison and Public Affairs

Current policy assigns full responsibility for legislative liaison and public affairs to the secretary of the Air Force. However, our analysis suggests that the secretary of the Air Force should allow the creation of distinct Space Force offices for legislative liaison and public affairs activities within the secretariat of the Air Force's offices. The secretary of the Air Force should give these distinct Space Force offices authority to advocate the positions of the Space Force independently of legislative liaison and public affairs efforts for the rest of DAF. Again, we seek this independence in response to the argument that a Space Force is being created, in part, to create and advocate positions that the Air Force has not been able to create in the past. Independent legislative liaison and public affairs offices could ensure that, once the Space Force creates new concepts and doctrine and proposes a new force structure, operational systems, and warfighting procedures, it has the authority to advocate these new ideas in Congress and in a broader public forum that could shape the decisions of politicians in Washington.

The secretariat should sustain such legislative liaison and public affairs offices within its own administrative organizations. The secretary of the Air Force should take a special interest in these offices to ensure that they are free to advance the substantive ideas that the Space Force is developing.

Support Activities

We did not conduct a detailed assessment of remaining support activities that an armed service would provide, as we did for the other functional areas. We assessed that the efficiencies of relying on many Air Force generic support services would likely outweigh any independence and sense of identity that self-provision might give the Space Force. These include base operations and security, nonoperational logistics and communications, and professional services, such as medical, dental, legal, and contracting services. That said, the Space Force would likely benefit from crafting formal agreements, such as MOAs, with the Air Force to clarify what services the Air Force should provide and agreements in which the Space Force potentially agrees to pay for some Air Force services.

As Chapter Five will highlight, personnel management will play an important role in ensuring that the Space Force maintains viable workforces. To be independent and have a sense of identity, the Space Force should have effective, substantive control over its recruiting and accession of personnel; appraisals, assignments, promotions, and retention of personnel; and training of personnel on space-specific topics. So long as it can maintain substantive control of decisions made and the information used to inform these decisions, the Space Force should rely on Air Force administrative programming, budgeting, financial, and personnel management processes to support these decisions. Using these processes should enhance efficiency and limit bureaucratic duplication.

Summary

We assessed which activities and organizations should transfer to the Space Force, using the performance elements of effectiveness, efficiency, independence, and identity. Our analysis shows that the majority of space organizations in the current services that are performing and directly supporting space operations and those supporting related training and acquisition should transfer to the Space Force. In its headquarters, the Space Force should have effective, substantive control over space-related concepts and doctrine, requirements and technical architectures, programming and budgeting, and communication with parties outside DoD, such as the White House, Congress, and the general public. Control of these activities, as well as of personnel management, is likely to be essential for the Space Force to be independent and have a sense of identity.

Although consolidating space activities under one new service could reduce duplications and seams, in certain areas it could also create new seams and gaps, potentially degrading integration of space with other terrestrial operations (air, sea, and land) unless these new seams are attended to. Addressing these new seams and gaps could have resource implications. The Air Force, for example, will need to develop its own space expertise to support Air Force space needs and establish liaisons and new relationships with the Space Force to access space-related expertise and services similar to what the Army and the Navy do today. The Space Force may also need to increase its representation at combatant commands to provide space support and its independent perspectives.

One space mission area that may stay fragmented across DoD is space control. Currently, many space control activities that the terrestrial services employ are framed as serving a supporting role to their operations (e.g., force protection). Lack of a coherent theory of space warfighting leaves many questions about which space control missions and capabilities should be controlled by the Space Force versus other services. This fragmentation of space control activities could hamper the Space Force's effort to build a common vision of space warfighting and to build an independent identity— and could ultimately undermine the value of establishing a separate service for space. The Space Force should make it a priority to develop and mature space warfighting concepts and doctrine to mitigate this risk.

CHAPTER FIVE

Career Field Sustainment Within the Space Force

One rationale for the creation of the Space Force as a separate service is the opportunity to have a fully dedicated cadre of space professionals who would devote their entire careers to developing and enhancing space capabilities.[1] The Space Force that will emerge from the transfer of activities and organizations, as discussed in Chapter Four, will be small in terms of personnel. This workforce would consist of members of the military and civilians with skills in space operations, space intelligence, space acquisition, and other STEM disciplines relevant to the Space Force's organize, train, and equip responsibilities. Many of these space professionals currently serve in the Air Force and the other armed services, and their career trajectories and the structure of their career fields are molded by those services. For transfers of space activities and organizations to be viable and to function as an independent service, the Space Force would have to be able to create and sustain its own career paths and career fields, and it would have to be able to grow and develop its own general officers.

However, given the expected size of the Space Force, relative to other services, the sustainment of healthy career fields could be challenging. A few career fields will likely be able to function autonomously within the Space Force, but the Space Force and the Air Force will likely need to share personnel in most Air Force Specialty Codes (AFSCs). Shared personnel could be workable within the Space Force if it could maintain a good amount of autonomy in its manpower and personnel processes via a robust manpower and personnel management function of its own, create its own personnel and manpower policies where necessary, and partner with rather than rely on the Air Force's manpower and personnel processes.

To investigate these issues, we conducted the following analyses. First, we developed a method to assess career field sustainability and applied it to key career fields critical for the functioning of the Space Force: 13S (Space Operations), 14N (Intelligence), and 62E/63A (Development Engineers and Acquisition Managers, which are managed together).[2] Because of the limited project timeline and access to personnel

---

[1] For example, see White House, 2019, p. 2.

[2] This chapter shows our analysis of the 13S and 14N career fields. Appendix B shows our analysis of the acquisition career fields.

data, we focused on officers and these career fields because of sponsor interest. That said, sustainability of the entire workforce associated with the Space Force should be evaluated. Second, we analyzed the capability of an independent Space Force to produce the required numbers of general officers to lead the service.

Determination of Space Force Officer Career Field Sustainability: RAND's Model

Given the personnel challenges inherent in the small size of the planned Space Force, we developed a model of sustainability for Space Force officer career fields, according to three criteria:

- **Pyramid health** incorporates both estimated numbers of members and promotion ratios for officers in the career field being examined. Projected promotion ratios are a good indicator of whether promotion timing and opportunities would be equitable between the Air Force and the Space Force, career field by career field. If Space Force promotion opportunities and timing are not equitable with the Line of the Air Force (LAF),[3] it would be a stretch to imagine airmen and members of other services willingly coming to the Space Force and staying within it.
- **Career path viability** examines whether enough variability in assignments—depth and breadth—within the Space Force exists to fully develop members over the course of their careers, utilizing the career field's talent management framework.
- **Senior leadership opportunities** examines the availability of O6 and command positions (squadron, group, and wing), as well as equivalent senior leadership opportunities in outside agencies (e.g., directorships in NRO, the Defense Intelligence Agency [DIA], or the National Security Agency [NSA]), as applicable.

It is most likely that the set of officer career fields critical to the Space Force would not fall neatly into "either-or" sustainability categories—that is, either fully sustainable within the Space Force or clearly not sustainable and permanently residing in the Air Force, with members on loan or on assignment to the Space Force. We would expect, instead, to see a continuum of outcomes between full sustainability and no sustainability. A simple continuum, for our current purposes, can be shown as including three categories of career fields (see Figure 5.1), based on assessment outcomes of the above three criteria:

---

[3] *LAF* refers to a grouping that includes all officers in operational or combat career fields and all officers in combat support career fields. LAF does not include medical specialties.

Figure 5.1
Career Fields Continuum

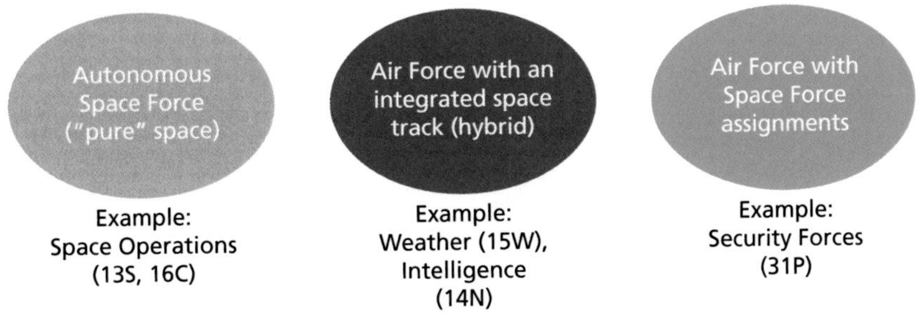

Autonomous Space Force ("pure" space)	Air Force with an integrated space track (hybrid)	Air Force with Space Force assignments
Example: Space Operations (13S, 16C)	Example: Weather (15W), Intelligence (14N)	Example: Security Forces (31P)

- **Autonomous Space Force:** This career field would be fully sustainable within the Space Force, or "pure" Space Force, with members having a full career within the Space Force alone. A presumed example of an autonomous Space Force career field would be 13S, Space Operations.
- **Air Force with an integrated space track:** In this career field, members remain within the Air Force, but with the understanding and accommodation that a substantive body of knowledge and experience about space is required. An analogous situation within the Air Force is the foreign area officer specialty. An officer undertakes special education to become qualified as a foreign area officer and then serves multiple assignments in this specialty before returning to the primary career field. Possible examples for the Space Force in this category of career field might be 15W (Weather) or 14N (Intelligence), both of which require specialized space knowledge, in addition to primary AFSC knowledge and experience, to be able to properly support the space mission. This category would likely be essential; even though several officer career fields would require high levels of specialized space knowledge to support the Space Force, they would likely be too small to be Space Force autonomous.
- **Air Force with Space Force assignments:** In this case, an officer remains in the Air Force, because his or her career field requires little specialized knowledge of the Space Force to be effective in a Space Force assignment or that knowledge can be quickly or easily acquired. An example of these career fields might be 31P (Security Forces). Security Forces officers could be assigned to the Space Force through the normal assignment process.

The three criteria—**pyramid health, career path viability, and senior leadership opportunities**—can be expressed in the form of a decision tree, as shown in Figure 5.2. The decision as to whether the career field is Space Force autonomous, Air Force with an integrated space track, or Air Force with Space Force assignments follows once the analysis of all three criteria has been completed for a specific career field.

Figure 5.2
Career Field Decision Model

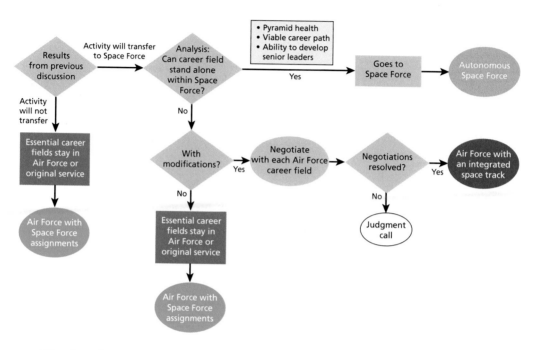

The first decision point in our decision tree refers to the results from the analyses reported in Chapter Four, which determined whether the career field *should* be part of the Space Force. Even if it would make sense to bring the career field into the Space Force, it is necessary to consider a follow-up question: *Can* this function or career field be moved into the Space Force? Can it work there as it is currently structured? Is it sustainable? The answer depends on the analytic outcomes with the three criteria: all three outcomes of the analysis of *pyramid health, viable career path,* and *senior leadership opportunities* must be positive for the model to yield the answer that a Space Force autonomous career field is possible. As can be seen in the model, a "yes" on all three is the vector that leads directly to the Space Force autonomous option.

If all outcomes are not positive, then moving to the parallel vector one level down leads to the question: Would modifications to the career trajectory ensure that specialized space knowledge can be acquired by these career field members? If the answer is "yes," negotiations must be conducted with that Air Force career field to determine the accommodations necessary for specialized Space Force knowledge. If negotiations are resolved favorably, the career field *can* be an Air Force career field with a special Space Force track. If negotiations are not fruitful, the fate of the career field is a judgment call for senior leadership.

If no modifications are desirable or possible, as seen in the downward vector, then the career field *cannot* be in Space Force as an autonomous career field, nor can it reside in the Air Force with a specialized space track. The career field will reside in the Air Force, and assignments to the Space Force will be part of the normal assignment process.

We will next demonstrate application of the three criteria (*pyramid health, career path viability,* and *senior leadership opportunities*) for four career fields of high interest to the Space Force: 13S (Space Operations), 14N (Intelligence), 62E (Development Engineers), and 63A (Acquisition Managers). We discuss 13S and 14N in this chapter; the two acquisition-related career fields are discussed in Appendix B. These findings, by necessity, are preliminary. A definitive analysis would require knowledge of the actual composition of the Space Force, which is of course not yet available.

The Case of 13S (Space Operations)

We assessed 13S (Space Operations) against the three criteria—*pyramid health, career path viability,* and *senior leadership opportunities*. If any career field within the Space Force would be expected to be autonomous, it would certainly be 13S. Further, if any career field within the Space Force *must* be autonomous, it would be 13S; it is difficult to imagine a credible rationale for a separate service otherwise.

Unfortunately, as it is now structured, 13S is not sustainable and would require structural changes to become so. In other words, a 13S officer would not be able to realize a full career in the Space Force. Sufficient senior leadership positions do not exist for 13S officers within the current billet structure. Moreover, as will be discussed more fully below, the structural deficiencies of the 13S career field compound the challenge of the Space Force generating the numbers of general officers it will need to lead a separate service. The Space Force's operators, the 13S career field, will not only need to generate enough general officers to lead its function but also need to generate a disproportionately high number of general officers to lead the Space Force more broadly.

Turning to the step-by-step 13S analysis, when *pyramid health* was examined for the 13S career field, the approximate 1,450 authorizations were the sufficient size.[4] The promotion ratios up through O5 would require minor adjustments to be equitable with LAF ratios (see Table 5.1), but those adjustments could likely be accomplished.

Career path viability for a Space Force 13S career field appears to be very feasible. A distinct advantage is that AFSPC has historically taken talent management seriously, and it has been out in front of the broader Air Force regarding specific selection criteria for its officers and in developing a space-specific developmental curriculum. The RAND team believes that career-focused feedback to officers at the five-year mark is a best practice. A new on-off, deploy-train cycle to enhance 13S expertise is a creative

---

[4] RAND's analysis included only those 13S officers within AFSPC. A follow-on analysis with all 13S officers in the Air Force, many of whom are assigned outside AFSPC, in institutional and other assignments, is recommended.

Table 5.1
13S Promotion Ratios

	Grade	Authorizations as of March 2019	Promotions	Selectivity to Higher Grade
Overall LAF	O1, O2, O3	25,227	O3 to O4	1.2
	O4	10,156	O4 to O5	1.6
	O5	7,449	O5 to O6	3.0
	O6	2,447	—	—
Overall 13S career field	O1, O2, O3	783	O3 to O4	1.1
	O4	367	O4 to O5	1.8
	O5	241	O5 to O6	4.7
	O6	51	—	—
Space Force 13S only	O1, O2, O3	653	O3 to O4	2.0
	O4	166	O4 to O5	2.5
	O5	81	O5 to O6	3.2
	O6	25	—	—

SOURCE: Data provided by the Air Force Personnel Center.

solution to the perennial Air Force challenge of finding time for training. AFSPC has been moving toward increased depth of expertise and experience for leadership, and the plan is for commanders to be assigned in their areas of depth.[5] In other words, much focused time and effort has already been expended to ensure a fully viable career trajectory for space operators, and the 13S career field would be moving over to the Space Force nearly intact.[6]

In terms of the availability of senior leadership and command positions and the question of whether the 13S career field within the Space Force would have enough of

---

[5]   We learned this during discussions with officers in the 13S career field.

[6]   More work is ongoing in preparation for a dedicated space cadre. Identification of key billets, analogous to key nuclear billets, is under way and is a major step forward in guaranteeing essential depth and experience in officers assigned to key positions in the Space Force. However, a key challenge for the 13S cadre in the Space Force would be the transition from a service orientation to four major warfighting functions. Two of the four compose the Air Force space mission set, but the other two would be new to the Space Force as it moves to include offensive capability. It would be a priority of the Space Force to select for, develop, and retain these new capabilities.

these positions to sustain the career field and provide leadership for the Space Force, the answer is a surprising no.

As seen in Table 5.2, analyses revealed that an **additional 27 O6 positions** would be needed to equal LAF O6 promotion rates, as shown in Table 5.2. (As discussed earlier, parity with LAF promotions is essential.) Twenty-seven additional O6 billets would require a major structural change to the current 13S career field.

Moreover, as the general officer ranks are considered, the outlook becomes worse. As operators of the Space Force, 13S officers would also be looked to for senior leadership of the overall Space Force, such as general officers required for CSO, vice CSO, combatant command staff, and joint staff representation. A restructuring of the 13S career field structure would help with general office throughput, but it would not address the challenge in its entirety.

An expanded 13S development pipeline would be needed to produce the general officers required. But the necessary career field structure and the leadership opportunities to build that pipeline simply are not there. The 27 additional O6 positions would help, but a deepening of both the O6 and O5 benches would be required. Promotions could be accelerated at the O5 and O6 grades, but then of course experience levels would suffer. Twenty-one additional general officer positions would be needed to equal LAF promotion rates. As a result of the sum of these issues, nearly one-half of the general officers in the Space Force will likely need to come from the Air Force or other services for the foreseeable future, as the fuller analysis reveals later in this chapter.

The sustainability outlook for space operators in the Space Force is summarized in Table 5.3. **This 13S structural issue and the challenge of general officer throughput in the Space Force are likely the two most serious results the RAND team encountered in the analyses described in this chapter.**

Table 5.2
Notional Authorization Options for Space Force

Grade	LAF Authorizations as of March 2019	Overall 13S Authorizations in Air Force	Notional Space Force Matching LAF	Difference
Company grade officer	25,227	783	804	+21
O4	10,156	367	323	−44
O5	7,449	241	237	−4
O6	2,447	51	78	+27
Total	45,279	1,442	1,442	—

SOURCE: Data provided by the Air Force Personnel Center.

Table 5.3
Sustainability of Autonomous Space Operators Career Field

Criteria	Sustainable?	Explanation
Pyramid health	Yes	The career field, at approximately 1,450 authorizations, would be of sufficient size. The promotion ratios up through O5 would require minor adjustments to be equitable with LAF ratios, but those adjustments could likely be accomplished.
Career path viability	Yes	The 13S career field would be moving to the Space Force nearly intact, and AFSPC has historically taken talent management seriously. Opportunities for depth, breadth, and leadership development at the necessary junctures have been established.
Senior leadership opportunities	No	At the O6 level, the structure of the Air Force 13S career field would have lower promotion rates (or higher phase points) than those of LAF and all other career fields. An additional 27 O6 positions would be needed to equal LAF promotion rates.
		At the general officer level, an expanded development pipeline would be needed to produce the general officers required. An additional 21 general officer positions would be needed to equal LAF general officer promotion rates.

The Case of 14N (Intelligence Officers)

Space intelligence officers require a good deal of space-specific knowledge, such as an understanding of orbital mechanics, to do their job well. Could the Space Force sustain the 14N career field autonomously? That is, could the Space Force access, educate, develop, and promote its own intelligence officers, not using a shared model with the Air Force but keeping its own Space Force intelligence officers for a full career within the Space Force alone?

Knowledge of which specific units will move to the Space Force is necessary to answer these questions. The analysis results in Chapter Four identified space intelligence personnel in space operational units, Air Force elements at NRO, NASIC, and ACC's 544 ISRG and 659 ISRG as those who should transfer, providing a coarse view of a space intelligence cadre that the Space Force might have. For the purpose of the career field sustainability analysis, however, additional details are needed: specific numbers of officers, ranks, and where they might serve (e.g., national-level intelligence community and new space intelligence units that might be established in the future), which are not fully known yet. Note that obtaining such data is more challenging for many career fields other than 13S.

As in the case of the 13S career field, we begin with an examination of *pyramid health*. We determined that a small number of space intelligence officers could produce a viable pyramid and that the promotion ratios are close enough to Air Force 14N promotion ratios to meet the demands of this first criterion (see Table 5.4).

For the question of whether the career field (if Space Force autonomous) has enough variability in both breadth and depth to grow and develop an intelligence officer over a full career, we examined the talent management framework of the career field to determine the answer. This framework, shown in Figure 5.3, is the type typically determined by the career field development team and career field and functional leadership. Our estimates suggest that the career field *would* have sufficient *career path variability* within the Space Force alone.

To determine the availability of *senior leadership opportunities*, we first examined the numbers against the overall structure of the career field and then, once again, examined the promotion ratios. When the availability of senior leader positions was analyzed for intelligence officers in our notional Space Force 14N career field, we did find sufficient numbers of O6 and command positions to sustain a full career path and provide leadership for the Space Force 14Ns, with one caveat: Two O6 directorships in agencies external to the Space Force, that would be equivalent to command positions, would need to be created. According to our subject-matter experts, these additions are highly feasible.

Table 5.4
Intelligence Career Field Promotion Ratios

	Grade	Authorizations as of March 2019	Promotions	Selectivity to Higher Grade
Overall LAF	O1, O2, O3	25,227	O3 to O4	1.2
	O4	10,156	O4 to O5	1.6
	O5	7,449	O5 to O6	3.0
	O6	2,447	—	—
Overall 14N career field	O1, O2, O3	1,848	O3 to O4	1.2
	O4	744	O4 to O5	2.3
	O5	387	O5 to O6	3.6
	O6	109	—	—
Space Force 14N only	O1, O2, O3	110	O3 to O4	2.1
	O4	26	O4 to O5	1.8
	O5	17	O5 to O6	4.3
	O6	4	—	—

SOURCE: Data provided by the Air Force Personnel Center.

Figure 5.3
Notional Intelligence Career Field Talent Management Framework

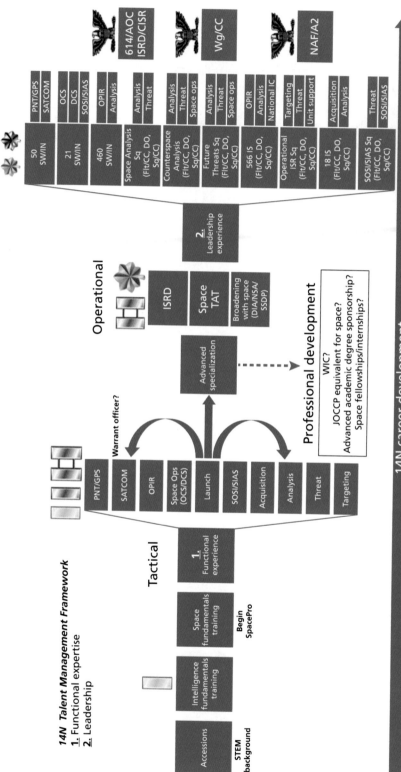

SOURCE: 548th ISRG, ACC.

NOTE: OPIR = Overhead Persistent Infrared; OCS = offensive counterspace; DCS = defensive counterspace; SOSI = space object surveillance and identification; SIAS = Sense Identify Attribute Share; WIC = weapons instructor course; JOCCP = Joint Officer Cryptologic Career Program; ISRD = Intelligence, Surveillance, Reconnaissance Division; TAT = tactical assessment team; SSDP = Space Security and Defense Program; SW/IN = space wing intelligence director; Flt/CC = flight commander; DO = director of operations; Sq/CC = squadron commander; IC = intelligence community; CISR = chief, ISRD; Wg/CC = wing commander; NAF/A2 = numbered Air Force intelligence directorate.

If these two command-equivalent positions, outside the Space Force, can be created, then our third criterion is met, and all three criteria will have been successfully addressed. We, therefore, have found that the 14N (Space Intelligence) career field could function as an *autonomous career field*. It would be small. The Space Force will need to heed the cautions that small career fields require more-deliberate management and are somewhat riskier—but it could be done.

A summary of our results for the 14N test case are in Table 5.5.

General Officer Throughput

In this section, we consider the numbers of general officers that the Space Force might have within its grade structure, based on different scenarios. We utilized a notional calculation of grade structures that matched the LAF grade structure ratios to make our assessments. Our analysis is not based on the number of organizations that will be part of the Space Force, nor does it address whether an organization would need general officer leadership or if O6 leadership would suffice.

According to data provided by the SFPTF, two scenarios for senior leadership are under consideration, lean and demanding. The lean Space Force would be led by 20 brigadier generals (O7s), 12 major generals (O8s), six lieutenant generals (O9s), and three full generals (O10s). The demanding Space Force replaces colonel leadership at a few locations with brigadier generals, increasing the number of O7s to 23 and O8s to 13, but leaves the most-senior leadership at six O9s and three O10s.

We utilized these numbers to build out the remaining grade structures with two different methods. In the first, labeled the *brigadier general ratio*, we used the LAF inventory grade ratio of O6s to O7s and applied that ratio to the lean and demand-

Table 5.5
Sustainability of Autonomous Intelligence Career Field

	Sustainable?	Explanation
Pyramid health	Yes	A preliminary look reveals a pyramid at least as robust as 14N in total, with the important caveat that small career fields must be carefully and deliberately managed.
Career path viability	Yes	The new talent management framework and the proposed ISR structure appear to offer the variability and development to sustain a small, highly specialized Space Force 14N. Results also support updated Air Force direction for increased intelligence specialization.
Senior leadership opportunities	Yes	There must be two additional O6 command-equivalent assignments in such agencies as NRO, DIA, and the Central Intelligence Agency (CIA) to achieve numbers, development, and quality of senior leaders and to avoid a decrease in Space Force O6 promotion rates compared with the Air Force's overall 14N and LAF rates.

ing number of O7s to determine how many O6s are needed to generate the number of brigadier generals under both scenarios. In our second method, labeled the *general officer ratio*, we used the LAF inventory ratio of O6s to the total number of general officers to determine how many O6s are needed under both the lean and the demanding scenarios.

We used LAF ratios because most of the officers who would move from the Air Force to the Space Force would come from LAF. LAF officers are accustomed to a particular set of management practices that govern promotion board timing, promotion opportunity to the different grades, and years of service to pin on ranks (phase points). If the grade structure in the Space Force were to vary dramatically from the Air Force's, this difference in grade structure could create pressure to promote different amounts of officers to the different field grades. For example, if there were three O5s for every O6 in LAF but five O5s for every O6 in the Space Force, the results would be downward pressure on the O6 promotion rate in the Space Force, as compared with LAF. If the LAF promotion opportunity to O6 (typically 60 percent) were applied to the Space Force, it would likely produce an excess of O6s in the Space Force, so the promotion opportunity to O6 would need to be lowered or the O6 phase point would need to be higher (i.e., slow down promotions). For these reasons, we believe that it would be prudent to align the Space Force grade ratios with those of LAF to the extent possible.

At the time of our analysis, we used the most-current LAF active duty inventory data (from the end of May 2019) to construct Table 5.6. From these data, we calculated the grade ratios we used to estimate the number of colonels for the two methods. First, using the brigadier general ratio method, there were 2,471 active duty O6s in the LAF inventory on May 30, 2019, and 138 brigadier generals, yielding a ratio of 17.9. Using that O6/O7 ratio with 20 brigadier generals in the lean Space Force yields 17.9 × 20 = 358 colonels. Similarly, with the 23 brigadier generals anticipated in the demanding Space Force, the structure would need 412 colonels to get the same support as seen in LAF.

Once we had our numbers of colonels, we used a similar approach to build the company and field grade structures. Starting with the LAF O5/O6 inventory ratio (7,864/2,471) and applying it to the 358 lean Space Force colonels yields 1,139 O5s, and so forth. Table 5.7 shows results for the lean Space Force and demanding Space Force grade structures that would match the LAF grade ratios from O1 to O7 using the brigadier general ratio method. Dividing the number of officers at any grade by the number of the next grade higher will yield a ratio in which the Space Force matches LAF, up through O7. (In both the lean and the demanding Space Force scenarios, we did not change the number of general officers in this force structure analysis.) The total number of officers derived in these grade structures was between 7,200 and 8,300.

For the general officer ratio, instead of using the O6/O7 LAF ratio to determine the number of colonels needed, we used the ratio of the number of O6s to the total number of general officers in LAF as seen in Table 5.6. In this case, there are 274 gen-

Table 5.6
Line of the Air Force Active Duty Inventory (May 2019)

Grade	Number of LAF Active Duty Officers
O1	6,535
O2	6,638
O3	15,680
O4	10,316
O5	7,864
O6	2,471
O7	138
O8	87
O9	36
O10	13

SOURCE: Data provided by the Air Force Personnel Center.

Table 5.7
Notional Inventory Options for Brigadier General Ratio Method

Grade	Number of LAF Active Duty Officers	Lean Space Force Based on 20 Brigadier Generals	Demanding Space Force Based on 23 Brigadier Generals
O1	6,535	947	1,089
O2	6,638	962	1,106
O3	15,680	2,272	2,613
O4	10,316	1,495	1,719
O5	7,864	1,140	1,311
O6	2,471	358	412
O7	138	20	23
O8	87	12	13
O9	36	6	6
O10	13	3	3
Total	49,778	7,215	8,295

SOURCE: Data provided by the Air Force Personnel Center.

eral officers in LAF, so the O6/general officer ratio is 2,471/274 = 9.02. Applying this ratio to the lean and demanding Space Force number of general officers (41 and 45, respectively) gives us 41 × 9.02 which rounds to 370 colonels for the lean Space Force and 45 × 9.2 which rounds to 406 colonels for the demanding Space Force. Using these numbers of colonels and applying the LAF grade ratios give the results in Table 5.8. In this method, the total number of officers ranged between 7,449 and 8,175.

Finally, the SFPTF expected the total number of Space Force active duty officers to be about 4,072, based on the *Space Cadre Quick Look*.[7] The above scenarios all greatly exceed that number of officers (providing a range of 7,200–8,300 officers). We, therefore, decided to estimate the grade structure that would match 4,072 total officers. Again, we applied the LAF grade ratios but instead built out a structure that matched all ratios from one grade to the next in which the total number of officers is limited to 4,072, including allowing the number of general officers to change to match the LAF grade structure. Table 5.9 shows the resulting grade structure: Instead of 41 or 45 general officers for the lean and the demanding Space Force, the total number of general officers generated by a structure that matches the LAF grade ratios is just 22, with 11 brigadier generals instead of 20 or 23.

Table 5.8

Notional Inventory Options for General Officers Ratio Method

Grade	Number of LAF Active Duty Officers	Lean Space Force Based on 41 General Officers	Demanding SF Based on 45 General Officers
O1	6,535	978	1,073
O2	6,638	993	1,090
O3	15,680	2,346	2,575
O4	10,316	1,544	1,694
O5	7,864	1,177	1,292
O6	2,471	370	406
O7	138	20	23
O8	87	12	13
O9	36	6	6
O10	13	3	3
Total	49,778	7,449	8,175

SOURCE: Data provided by the Air Force Personnel Center.

---

[7] Barry Royx, *Space Cadre Quick Look*, 3rd ed., Peterson Air Force Base, Colo.: Air Force Space Command, February 2019, internal document.

Table 5.9
Notional Inventory for 4,072 Space Force Officers

Grade	Number of LAF Active Duty Officers	Space Force Limited to 4,072 Officers
O1	6,535	535
O2	6,638	543
O3	15,680	1,283
O4	10,316	844
O5	7,864	643
O6	2,471	202
O7	138	11
O8	87	7
O9	36	3
O10	13	1
Total	49,778	4,072

SOURCE: Data provided by the Air Force Personnel Center.

Given our initial assumption that it will be desirable (for recruiting, retention, and morale) to structure the Space Force so that its promotion management practices are equitable with the Air Force's, we reached the following conclusions based on the above analysis:

1. With a Space Force grade structure similar to the Air Force's, and given the current, projected overall numbers of officers within the Space Force, the Space Force would *not* be able to organically produce the numbers of general officers it needs under either the lean or demanding models.
2. For the foreseeable future, the Space Force will likely need to draw about one-half of its general officers from the Air Force and other services.

Challenges

Because the 13S structure can be remedied, the 13S (Space Operations) and the 14N (Intelligence Officers) career fields are perhaps the most likely of a very small number of uniformed career fields to be able to support full careers in the Space Force. Most career fields will, by necessity, be shared with the Air Force.

Career fields can be decked fairly easily across mission sets of the Air Force, but there is one precedent wherein a group of AFSCs must accommodate a mission set with very different and highly controlled requirements. That is the case of the conventional-nuclear split that many career fields within the nuclear enterprise must navigate. The nuclear enterprise has only two all-nuclear career fields in which airmen spend their full careers: the officer career field of 13N (Nuclear and Missile Operations) and enlisted career field of 2W2 (Nuclear Weapons Maintenance). Therefore, personnel in all other nuclear enterprise career fields toggle between conventional and nuclear proficiencies and assignments, and this approach is how the Air Force maintains its dual capability. There are great challenges in implementing dual capability, however.

The nuclear enterprise continues to improve on its dual capability model at the career field level, but it took several years even beyond the stand-up of Global Strike Command to refine the approach and overcome serious deficiencies in several career fields that plagued the nuclear enterprise from pre-2007 to about 2012.[8] What makes this shared approach so challenging is that, within the nuclear enterprise, each shared career field must accommodate the movement between conventional and nuclear in its own unique way. What works for one career field does not work for another; no two are alike.

Ultimately, greater flexibility in personnel policies and processes and subsequent policy changes have allowed dual capability within the Air Force to be strengthened.[9] This lesson will be an important one for a successful partnership between the Air Force and the Space Force.

---

[8] Shirley M. Ross, *Final Report: Nuclear Enterprise Human Capital Strategy and Execution Progress, 2010–2012*, Washington, D.C.: Deputy Chief of Staff of the Air Force for Manpower, Personnel, and Services, October 15, 2012, internal document.

[9] Many organizations aided the reinvigoration of the nuclear enterprise by changing existing or creating new personnel and education policy and by demonstrating flexibility and innovation in policies and processes when critically needed. The results of those efforts speak to the amount of progress that can be made when organizations recognize the need for flexibility in current policies or the need for new policies. Among the most important policy changes were the creation of the enlisted development teams, championed by then–Chief Master Sergeant of the United States Air Force James Roy, the Human Capital Strategy Team, and the Air Force Directorate of Force Development, which enabled the nuclear enlisted career fields to select the most qualified senior noncommissioned officers and to plan for development of senior enlisted leaders more deliberately (this innovation later expanded across the Air Force); the prescreening for Personnel Reliability Program (PRP) certification instituted by AETC, which saved many training dollars and increased accession pipeline efficiencies; and the initial process work with Air Force Personnel Center (AFPC) that led directly in later years to the AFPC Centralized PRP Cell pilot and implementation, which had and continues to have far-reaching impacts. Examples of a flexible approach that aided the reinvigoration are the accelerated manpower studies schedule to enable the rewrite of nuclear enterprise guidelines that were more than a decade out of date, by AF/A1M (Manpower) and AFPC/MA (Manpower Directorate), and AETC's creation of nuclear education courses 200, 300, and 400 and its addition of nuclear education curricula in professional military education. For a full discussion of new policies and demonstrated flexibility, see Ross, 2012, pp. 74–110.

Implications for the Space Force

It cannot be assumed, therefore, that airmen or Space Force personnel could move seamlessly into and out of the Space Force or that a shared model will necessarily be easy to establish and maintain over time. A successful shared approach will require a great deal of deliberate planning and implementation to construct. If the inevitable challenges with a shared model are ignored or not taken seriously, the lessons of history will be repeated: Personnel will not be able to perform to standards, the career fields themselves will experience serious issues requiring time-consuming remediation, and overall performance of the service will suffer.

And even where the Space Force can sustain career fields autonomously, challenges will remain. The Air Force has historically struggled with small career fields. First, personnel processes and policies that work well for large numbers often do not allow for the adroitness needed to manage small numbers with the required subtlety and precision. Second, a major, unforeseen event, in the closed personnel systems of the services, can turn disastrous without large numbers to fall back on for resiliency. A recent example is the cheating scandal at missile bases in Air Force Global Strike Command, in which significant numbers of officers were lost.[10] Consequently, the Air Force's 13N (Nuclear and Missile Operations) career field has no way to recover the numbers of officers now missing in the particular year groups that were involved.

One important consideration for the design of the Space Force and its personnel systems is the success of several specialized missions that have lobbied for and received some independence to build and run their own personnel processes. Intelligence and special operations are good examples of entities with high autonomy in their personnel processes, resulting in personnel processes that work more effectively for their missions. AFSPC has already been successful in determining its own selection standards (STEM degrees) and in mounting its own developmental programs.

New authorities granted to the service secretaries in the 2019 NDAA language may also provide potential benefits to the Space Force. Sections 501 and 502 in combination, for example, will allow the Space Force to bring in older, more-experienced personnel from the private sector with particular technical talents and grant them "constructive credit," which essentially amounts to their coming in as a captain, major, or even lieutenant colonel, provided that they have those levels of skills and abilities. Because the Space Force will not be operationally deploying to war zones in combat roles, the feasibility of recruiting and retaining older, highly experienced professionals may be greater for the Space Force than for other services.[11]

---

10 Ernesto Londoño, "US Air Force Relieves Nine Officers Following Nuclear Cheating Probe," *Washington Post*, March 27, 2014.

11 Pub. L. 115-232, John S. McCain National Defense Authorization Act for Fiscal Year 2019, Title V, Military Personnel Policy; Subtitle A, Officer Personnel Policy, Sections 501–502, August 13, 2018.

Conclusions

Overall, after examination of the issues covered in this chapter, the best outlook for career field sustainability for the Space Force may consist of constructing its own small set of autonomous or stand-alone Space Force career fields, along with a larger set of career fields shared with the Air Force, through the use of a four-pronged approach:

1. Use careful analyses to determine which career fields are candidates for autonomous, or stand-alone, Space Force career fields and stand them up.
2. When considering shared career fields with Air Force, pay attention to lessons learned from the Air Force dual-capability, conventional-nuclear model and the specific approaches that vary by each Air Force nuclear career field.
3. Leverage the new flexibilities as allowed by the recent NDAA language.
4. Maintain a substantial level of Space Force independence in personnel, manpower, education, and training processes to better craft these vital functions for a small service.

Small and highly specialized, the Space Force will not be able to function, sustain its career fields, and perform well using standard Air Force policies designed for much larger forces. But careful attention to forging its own path for career field sustainability and being realistic about the numbers of general officers that can be generated will greatly assist, as will being open to new ideas and new ways of imagining form and function for an innovative and successful future Space Force.

CHAPTER SIX

Lessons from the Creation of Other Military Organizations

As part of our support for the SFPTF, we examined other instances in which DoD created new organizations to learn more about why DoD turns to organizational change to address policy problems, to consider how these organizations evolved, and to set the broader context to consider how the Space Force will compare with other organizations within DoD. The team considered nine organizations: an independent service (the Air Force), services and branches inside a military department (the Marine Corps, the Army Air Corps, and, to some extent, the Coast Guard), combatant commands (CYBERCOM and SOCOM), specialized organizations (NRO and MDA), and even an Air Force major command (Global Strike Command).

We discuss each organization in detail in Appendix A. Here, we consider them together to draw cross-cutting implications for the Space Force. Since we consider effectiveness, efficiency, independence, and identity to be critical for an organization's performance, we look at the role of each factor for these organizations to draw out potential implications for the Space Force. In particular, we describe what each of these factors might add, along with potential challenges. To complete our comparative work, we also consider how the size of the Space Force compares with the size of other organizations.

Effectiveness

To be successful, the Space Force must seek to improve the *effectiveness* of the U.S. armed forces. Given the emerging threats to U.S. space-based assets, this is no small task. In Chapter Three, we define *effectiveness* as "the level of military capability that the joint force can achieve within a fixed level of resources." The nine organizations we examined were established to develop military capabilities in a new domain, serve a particular policy function, or solve a perceived problem. For example, the Army Air Corps, the Air Force, and CYBERCOM were all to exploit new warfighting domains. The Marine Corps and the Coast Guard were both established to fulfill key missions for the emerging U.S. nation. The Marine Corps was originally created to provide shipboard security and to have the capability to board adversary ships. The Coast Guard was created to enforce tariffs and customs. MDA and SOCOM were both organized around building high-end military capabilities (missile defense and elite special operations). NRO was established to provide better intelligence against the Soviet Union, and Global Strike Command was instituted to address shortfalls in the military's handling of the nuclear arsenal.

Each of these organizations continues to exist, with the exception of the Army Air Corps, which, of course, evolved into the Air Force. Cynics might say that this is due to bureaucratic inertia, but there is an argument to be made that each organization has been able to demonstrate some level of increased effectiveness. Over time, critics of these organizations have raised questions about whether there is a continued need for their existence. As we note in Appendix A, following World War II, the Marine Corps worked to establish itself as a Title 10 entity to rebuff criticism that it merely duplicated capabilities provided by the Army. SOCOM is currently working to develop relationships with partner and allied military forces, as it pivots from its focus on counterterror operations.[1] Similarly, the Space Force will need to demonstrate effectiveness and show how it can be effective as geopolitical, environmental, technological, and other factors change. Otherwise, the Space Force will likely face calls for its reintegration into other parts of DoD.

One recurring issue we note in our cases has to do with whether an organization directly engages in combat or if it works primarily to enable others. To be sure, providing logistical support, such as fuel or ammunition, or even providing acquisition or legal support, can play a critical role in reaching national objectives. Nevertheless, inside DoD, there is prestige associated with being a "warrior" that can affect intradepartmental competition for resources or favorable policy rulings. Occasionally, DoD officials express concern about the "tooth-to-tail ratio," or the proportion of combat forces to those supporting them.[2]

As the case study of the evolution of the Air Force illustrates, powered aircraft were initially used to provide reconnaissance and communications, but the development of pursuit aviation and, later, strategic bombing gave the fledgling Army Air Corps the justification it needed to become an independent military service. Being able to directly unleash violence against an adversary helps provide an identity and prestige to a military service. The Marine Corps has been able to build an image of itself around warfighting, and the Coast Guard directly employs ships to interdict drug

---

[1] Jen Judson, "Where Do Special Forces Fit in the National Defense Strategy?" *Defense News*, May 17, 2019.

[2] For an attempt to measure tooth to tail over time, see John J. McGrath, *The Other End of the Spear: The Tooth-to-Tail Ratio (T3R) in Modern Military Operations*, Fort Leavenworth, Kan.: Combat Studies Institute Press, 2007. On efforts to increase the efficiency of tail forces, see Jacques S. Gansler and William Lucyshyn, *Improving the DoD's Tooth-to Tail Ratio*, College Park: Center for Public Policy and Private Expertise, University of Maryland, February 2014.

smugglers. It remains unclear whether and how much the Space Force will be able to play a role as an employer of force rather than just as an enabler of others that do so.[3]

Efficiency

An organization demonstrates improved *efficiency* if it uses fewer resources to achieve a particular level of capability than was required in the past. Some of the justification for creating the Space Force is that it will be able to consolidate military space assets and functions in one place, which will allow it to realize economies of scale. Similar motivation led to the creation of the Coast Guard (to centralize authority over maritime safety and security), CYBERCOM (to unify the military cyber enterprise), and NRO (to reduce duplication of effort in gathering intelligence across the services).

This motivation is not always fulfilled. Our cases indicate that, when DoD creates a new agency or service, it can create additional cleavages across the greater defense enterprise. For instance, MDA was created to help DoD with missile defense, but its separation from the Army's operations led to the design and acquisition of systems that the Army cannot use well. This is best displayed by Terminal High Altitude Area Defense (THAAD)—which some have criticized for the time it takes to deploy it to an operating location.[4] Similarly, NRO was created to help mature space capabilities and provide imagery intelligence during the Cold War. However, during the Gulf War, there were complaints about NRO's lack of responsiveness to operators' requests.[5] To the extent that the Space Force will focus on missions that will be important to it, such as space control, and not direct support of other services, the Space Force could face similar criticism.

Independence

Advocates for creating a Space Force argue that it is necessary to have an autonomous organization that can focus solely on the space domain.[6] Instead of competing with combat air forces, mobility air forces, and other priorities, an independent Space Force could instead allocate its resources toward space alone.

---

[3] SPD-4 explicitly notes that "[n]o longer do space systems simply enable terrestrial forces to fight and win wars; actions in space also will directly contribute to the outcomes of future crises or conflicts" (White House, 2019, p. 1).

[4] This is based on our conversations with European Command staff.

[5] Clayton D. Laurie, *Congress and the National Reconnaissance Office*, Washington, D.C.: Office of the Historian, National Reconnaissance Office, June 2001, p. 7.

[6] For example, see Matthew Donovan, "Unleashing the Power of Space: The Case for a Separate US Space Force," *War on the Rocks*, August 1, 2019; and Harrison, 2018b.

Budget independence is vital for the successful establishment of a service. Budgetary control allows an organization to manage key functions, such as personnel, promotion, acquisition, and operations. The Marine Corps, the Air Force, and SOCOM all have independent budgets, allowing them to create and pursue their own agendas. The Coast Guard's Title 10 authorities protect the organization from other law enforcement agencies. Major Force Program 11 (MFP-11) allows SOCOM to act almost as if it were a service of its own, capable of independently acquiring equipment and funding operations. In contrast, CYBERCOM and Global Strike Command lack the same kind of budgetary independence, leading to reliance on others, which affects their ability to set priorities on their own.

Lack of independence can hinder an organization. For instance, despite being a unified command, CYBERCOM has been unable to gain control of its service-specific components.[7] Each service maintains its own network, which complicates CYBERCOM's ability to defend them. Despite its service-like status because of its Title 10 authorities, SOCOM displays a tense relationship with the other services over control of resources, missions, and personnel. NRO's reliance on the services and the intelligence community to provide personnel has led to chain-of-command conflicts and has contributed to budget battles with other organizations.

The more the Space Force potentially relies on others, the more risk it incurs in being a service in name only. As we argue below, at its currently proposed size, the Space Force will have to rely on the secretariat and Air Force for a wide range of functions. The small size of the Space Force relative to other military services could pose difficulties. Chapter Five considers how the small size of the Space Force will affect its ability to field a core of general officers, as well as its officer corps in key career fields.

Identity

To fulfill the charge of SPD-4, and to continue to exist in the future, the Space Force will need to establish its own identity.[8] Our examination of other organizations shows that creating an identity helps an organization distinguish itself from others, provides a vision to its own members, and helps generate external support.

The creation of the Air Force as an independent service and the solidification of the Marine Corps as a distinct force were both largely based on their ability to contribute to military operations and to develop warfighting doctrine. Both the Army Air

---

[7] DoD defines a *unified command* as a "command with a broad continuing mission under a single commander and composed of significant assigned components of two or more Military Departments that is established and so designated by the President, through the Secretary of Defense with the advice and assistance of the Chairman of the Joint Chiefs of Staff." See Office of the Chairman of the Joint Chiefs of Staff, *DoD Dictionary of Military and Associated Terms*, Washington, D.C., October 2019, p. 224.

[8] White House, 2019.

Corps and the Marine Corps made key contributions toward victory in World War II. The crucial roles played by amphibious warfare in the Pacific Theater and airpower in both theaters fostered widespread support for both services. At the same time, both services crafted and advocated doctrine centered on bringing unique and critical, and arguably stand-alone, warfighting capabilities to DoD.

Space capabilities have also contributed to military operations, most notably in the period since the Gulf War of 1991. Nevertheless, these contributions have been limited to an enabling role. The lack of space warfighting doctrine has denied the space community a common vision and has contributed to the existence of separate communities within the space enterprise.

Size

One issue came up during our analysis that we should consider before proceeding. The size of the Space Force could pose a serious obstacle to its success. DoD is attempting to limit the additional resources needed to build the Space Force, which is understandable and even laudable, but being small could hurt the viability of the Space Force. To be sure, having a large budget or a large number of people in an organization does not guarantee, or even necessarily increase the likelihood of, organizational success. Nevertheless, to some extent, having more resources allows an organization to have more power over outcomes and more of an ability to apply those resources toward fulfilling its mission. In addition, DoD leaders do compare themselves with one another according to how much budget they control and how many people they lead.

For our look at organizations within DoD, we considered size in terms of overall budget and end strength. We examined the budgets of several different organizations in DoD and compared them with the proposed $15 billion that the Space Force will control.[9] As shown in Figure 6.1, the Space Force would have more annual spending power than many other organizations but considerably less than the Air Force had on its inception and less than the Marine Corps. For a service that oversees sophisticated and expensive technology, such as satellites, a relatively large budget would seem to be appropriate.

As it is currently conceived, the end strength for the Space Force is quite low. At 16,000 personnel, the Space Force would be dwarfed by other services. On its inception in 1947, DAF included 300,000 people, and the Marine Corps today consists of just over 186,000 (see Figure 6.2). The Space Force will be almost 39 percent of the size of the Coast Guard, which has around 40,557 serving on active duty.

---

[9]   We limit our discussion to unclassified budget data. For the 1926 Air Corps and 1947 Air Force budgets, we represent their budgets in inflation-adjusted dollars to make them comparable with current budgets.

Figure 6.1
Budgets

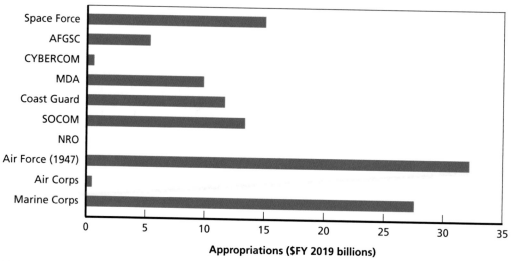

Appropriations ($FY 2019 billions)

SOURCES: For the Marine Corps, see Department of the Navy, Office of Budget, *Highlight of the Department of the Navy FY2019 Budget*, Washington, D.C., 2018, p. 8. For the Air Corps and Air Force (1947), see Rudolf Modley and Thomas J. Cawley, *Aviation Facts and Figures, 1953*, Washington, D.C.: Lincoln Press, 1953, pp. 102–103, 106. For SOCOM, see Congressional Research Service, *U.S. Special Operations Forces (SOF): Background and Issues for Congress*, Washington, D.C., March 2019, p. 7. For the Coast Guard, see U.S. Coast Guard, "U.S. Coast Guard Fact Sheet: Fiscal Year 2019 President's Budget," Washington, D.C., 2018. For MDA, see MDA, *Fiscal Year (FY) 2019 Budget Estimates: Overview*, Fort Belvoir, Va., 2018, p. 1. For CYBERCOM, see U.S. Senate Armed Services Committee, statement of General Paul M. Nakasone, Commander, United States Cyber Command, before the Senate Committee On Armed Services, Washington, D.C., February 14, 2019. For AFGSC, see James M. Kowalski, *Air Force Global Strike Command*, briefing, Barksdale Air Force Base, La., May 2013, p. 26. We omitted NRO because of a lack of unclassified information about its budget.

Even when we turn to nonservice organizations, the Space Force will have significantly fewer people than SOCOM and have fewer than one-half the forces assigned to Global Strike Command, a major command within the Air Force.

One argument in favor of a smaller force is that the Space Force will emphasize technology and will not need to have as many people as the Army or Marine Corps do, both of which need to populate large units. Although it is difficult, and perhaps unfair, to compare the Space Force with these services based on end strength, it is important to consider how the Space Force will be considered within the U.S. armed forces. The small number of people in the Space Force might make it difficult to create a service culture akin to that found in other services.

Ultimately, the ideal size of the Space Force will likely correspond to its assigned missions and responsibilities. If its mission is similar to that of space forces within the armed forces today, it might not need much more than 16,000 people to carry that

Figure 6.2
End Strength

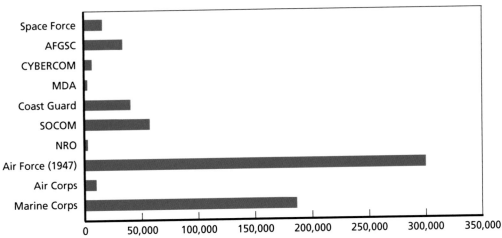

SOURCES: For the Marine Corps, see Senate Armed Services Committee, *FY20 National Defense Authorization Act Executive Summary*, Washington, D.C., 2019, p. 13; and Defense Manpower Data Center, "Active Duty Military Personnel by Service by Rank/Grade," data set, updated September 30, 2019. For the Coast Guard, see Department of Homeland Security, *Department of Homeland Security: U.S. Coast Guard Budget Overview Fiscal Year 2020 Congressional Justification*, Washington, D.C., 2019, p. 37. For SOCOM, see Congressional Research Service, *Defense Primer: Special Operations Forces*, Washington, D.C., 2018, p. 1. For CYBERCOM, see CYBERCOM Public Affairs, "Cyber Mission Force Achieves Full Operational Capability," CYBERCOM, May 17, 2018. For AFGSC, see DAF, *Air Force Global Strike Command Strategic Plan 2018*, Washington, D.C., 2018, p. 10. For MDA, see MDA, *Notification and Federal Employee Antidiscrimination and Retaliation Act (NoFEAR): Fiscal Year 2018 Report*, Washington, D.C., 2019, p. 1. For NRO, see NRO, "About the NRO: Who We Are," webpage, undated. For the Air Corps and Air Force (1947), see Modley and Cawley, 1953, pp. 102–103, 106.

mission out. If its mission and associated responsibilities increase, however, it might need to increase its end strength commensurate with those additional missions. Few would argue that the Space Force should be assigned more people just to allow it to stand on equal footing with other services or to be able to develop its own culture. If the Space Force's mission grows, its end strength will likely grow as well.

CHAPTER SEVEN

Charting an Adaptive Approach to Implementation

Chapter Six summarized examples of DoD's experience with reorganizing military units. As the case studies show, comprehensive planning is essential, but even the best of plans cannot anticipate all the contingencies that may arise, whether through alterations in expected funding, changes in leadership, or surprises in the threat environment. This chapter presents a high-level view of how Space Force planning might successfully navigate the expected—and unexpected—challenges of implementation. In approaching this chapter, we examined the literature of decisionmaking under conditions of deep uncertainty and adaptive planning and then considered how to apply these approaches to implementation of the Space Force.[1] Our aim is to offer guidance on how to successfully grow the Space Force while coping with a range of potential vulnerabilities in its plans. The Space Force will need to be adept at anticipating warning signs and identifying actions along the way to mitigate impacts of "unplanned" future conditions and thus increase the likelihood of successful implementation across a range of possible futures, with an eye toward both growth and resilience in the face of uncertainties.

The objective of the Space Force is to "to ensure unfettered access to, and freedom to operate in, space, and to provide vital capabilities to joint and coalition forces in peacetime and across the spectrum of conflict" (ends) and to do so by outpacing future threats, institutionalizing advocacy of space priorities, and further the building of a space warfighting culture (means).[2] Achieving these objectives hinges on many assumptions about how the threat environment will continue to evolve; how the Space

---

[1] For deep uncertainty, see Vincent A. W. J. Marchau, Warren E. Walker, Pieter J. T. M. Bloemen, and Steven W. Popper, eds., *Decision Making Under Deep Uncertainty: From Theory to Practice*, Cham, Switzerland: Springer, 2019. For adaptive planning, see James A. Dewar, Carl H. Builder, William M. Hix, and Morlie Levin, *Assumption-Based Planning: A Planning Tool for Very Uncertain Times*, Santa Monica, Calif.: RAND Corporation, MR-114-A, 1993; Robert J. Lempert, Steven W. Popper, and Steven C. Bankes, "Confronting Surprise," *Social Science Computer Review*, Vol. 20, No. 4, Winter 2002; Marjolijn Haasnoot, Jan H. Kwakkel, Warren E. Walker, and Judith der Maat, "Dynamic Adaptive Policy Pathways: A Method for Crafting Robust Decisions for a Deeply Uncertain World," *Global Environmental Change*, Vol. 23, 2013.

[2] White House, 2019, pp. 1–2. For an argument about the need to create a separate culture for space, see John Galer, "Only a Separate Service Can Create a Space Culture," *SpaceNews*, June 9, 2019.

Force will be resourced and operated within the current and future Air Force and DoD; how its staff will be recruited, developed, managed, and retained; and how rapidly space-focused CONOPs and supporting technologies can be deployed. Some of these factors cannot be reliably predicted. For this reason, the Space Force would benefit from an approach to planning that aims for robustness across a range of futures rather than optimality for a single (hoped-for) future.

The chapter is organized as follows. After first reviewing the basic principles of adaptive planning, we step through the vulnerabilities of the recommended initial operating capability (IOC), organized according to the functions addressed earlier in this report: operations, acquisitions, career management, and headquarters functions, such as requirements, PPBES, and public and legislative affairs. In each case, we examine the underlying assumptions of the IOC and consider how the success of the IOC and ultimately the full operational capability (FOC) could be undermined if one or more critical assumptions prove to be weak or invalid in the face of future events.

Adaptive Planning

Adaptive planning is a simple concept: Conduct planning with the knowledge that the future is uncertain, key planning assumptions may prove to be invalid over time, and more and better information can guide adjustments and redirection in plans and their implementation.[3] Paul Davis broadened the term *adaptive planning* to include flexibility, adaptation, and robustness, which he calls the "FARness principle."[4] The idea is for planners to ensure that decisionmakers have a rich set of options from which to choose, thus raising the likelihood of successful implementation over a range of possible futures. Put another way, Jan Kwakkel explains:

> Adaptive planning means that plans are designed from the outset to be altered over time in response to how the future actually unfolds. In this way, modifications are planned for, rather than taking place in an ad hoc manner. The flexibility of adaptive plans is a key means of achieving decision robustness.

> This means that a wide variety of futures has to be explored. Insight is needed into which actions are best suited to which futures, as well as what signals from the unfolding future can be monitored in order to ensure the timely implementation

---

[3]   Lempert, Popper, and Bankes, 2002.

[4]   Paul K. Davis, *Analysis to Inform Defense Planning Despite Austerity*, Santa Monica, Calif.: RAND Corporation, RR-482-OSD, 2014, pp. 8–11.

of the appropriate actions. Adaptive planning thus involves a paradigm shift from planning in time, to planning conditional on observed developments.[5]

In any construct of adaptive planning, the key challenges for planners are to know what signals to look for and when and how to act to change course and avoid failure. Failure is expressed in terms of not meeting strategic objectives. An early variant of adaptive planning is assumption-based planning, originally developed for the Army at RAND and widely applied to military and other government planning activities.[6] Using the terminology introduced by James and coauthors and shown in Figure 7.1, planners first identify the "important assumptions" that underlie their plan or strategy (step 1).[7]

Assumptions can be implicit or explicit. An example of an explicit assumption is that the IOC specifically states that the Space Force will function within DAF, just as the Marine Corps functions as a quasi-independent component within the Department of the Navy. An example of an implicit assumption for the IOC—that is, unwritten in the IOC itself—is that the defense budget will fully resource the IOC to achieve the FOC by 2021. Making implicit assumptions explicit before plans are locked in

Figure 7.1
Steps in Assumption-Based Planning

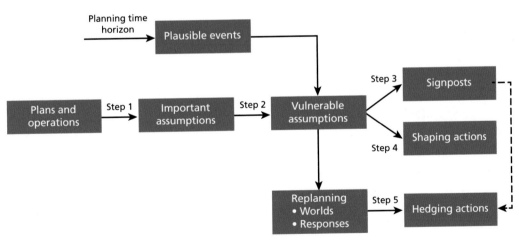

SOURCE: Dewar et al., 1993, Figure 7.1.

---

5   Jan Kwakkel, "Managing Deep Uncertainty: Exploratory Modeling, Adaptive Plans, and Joint Sense Making," *Integration and Implementation Insights*, August 15, 2017.

6   Dewar et al., 1993; James A. Dewar, *Assumption-Based Planning: A Tool for Reducing Avoidable Surprises*, New York: Cambridge University Press, 2002.

7   Dewar et al., 1993.

place is an often-neglected step in military (and other) planning.[8] Failing to make important assumptions explicit and failing to adequately plan for contingencies carry potentially serious implications for successful implementation.

To implement this approach, planners would begin by generating a list of assumptions and then distinguishing among those that their plan's success vitally depends on from those that are important but less critical. Following the logical sequence in Figure 7.1, from the initial list of assumptions, planners and other stakeholders determine which of these are "load-bearing," or "vulnerable assumptions," over the course of the planning time horizon (step 2). A load-bearing assumption is one that, if broken by one or more plausible future events or set of conditions, could undermine the success of a plan or strategy or, even worse, cause the plan to fail entirely. For example, insufficient personnel in selected career tracks could substantially undermine the sustainability of necessary expertise and leadership within the Space Force. As another example, the inability of the Space Force to advocate for its budget could result in curtailed or canceled acquisitions that undermine newly developed space-based CONOPs.

The point of identifying vulnerable assumptions is to monitor changes in the internal and external environment (step 3, "signposts") that indicate when such prospective vulnerabilities become real and then devise actions (step 4, "shaping actions") that could help to forestall or mitigate the realization of the vulnerability. For example, signposts for the IOC could be in the form of budgetary decisions that fail to align with the assumptions embedded in the IOC.

Triggered by a signpost, a consequent shaping action could be the immediate elevation of concerns about the budget prior to congressional enactment to avoid the deleterious effects of a misaligned budget in the critical first year of the IOC. In some cases, a shaping action might not be available or had been tried and failed, in which case, planners must reconsider their plans in substantial ways and develop "hedging actions" (step 5) and thus provide an alternative approach to meeting the plan's objectives. For example, a hedging action could lead to a scaling back of the timeline for implementing a new CONOP or reliance on less expensive technologies to achieve desired effects. As Dewar and coauthors note, step 5 is the most time- and labor-intensive step and "the epitome of planning in that it contemplates worlds that are unexpected or undesirable."[9]

Another well-established form of adaptive planning is called dynamic adaptive planning pathways (DAPP). DAPP explicitly accounts for contingencies affecting an

[8]   Dewar et al., 1993; Lempert, Popper, and Bankes, 2002; Dewar, 2002; Robert J. Lempert, Horacio R. Trujillo, David Aaron, James A. Dewar, Sandra H. Berry, and Steven W. Popper, *Comparing Alternative U.S. Counterterrorism Strategies: Can Assumption-Based Planning Help Elevate the Debate?* Santa Monica, Calif.: RAND Corporation, DB-548-RC, 2008; Davis, 2014.

[9]   Dewar et al., 1993, p. 50.

initial plan that may occur over different time horizons.[10] DAPP follows a similar thought process as assumption-based planning but does so in an explicitly iterative manner over the course of the planning horizon, as shown in Figure 7.2. For the Space Force, the time horizon could extend beyond its start-up phase of the IOC and its targeted FOC to include two or more Future Years Defense Programs. Some actions may work well for some period and over some set of futures but then may reach tipping points or points of diminishing returns in the face of future events. Space Force leadership will want to remain alert to changing future conditions and adjust plans accordingly to remain on a productive and sustainable path. Alternative paths would of course carry their own costs and risks, thus presenting decisionmakers with trade-offs to consider across operational and other planning objectives.

These basic concepts of adaptive planning provide a structure for considering issues that may arise in the implementation of the IOC and, later, the FOC, but also provide a structured approach to thinking about alternative implementation paths over a longer period. Given the rapid pace of space-based technological development and the emerging threats from adversaries, the Space Force's planning process could consider monitoring changes in both the internal and external environments in a timely way, ensuring that alternative pathways were always at the ready, and deciding when the time would be right to change course. Thus, adaptive planning can also be agile,

Figure 7.2
A Notional Example of Dynamic Adaptive Planning Pathways

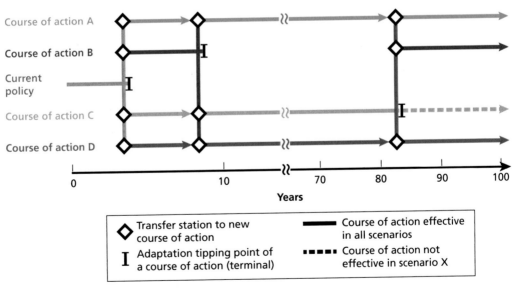

SOURCE: Haasnoot et al., 2013, Figure 2 (CC BY-NC-ND 3.0).

---

[10] Haasnoot et al., 2013.

where the pacing of threats, technological developments, and budgetary shifts determine the pivot points in the planning cycle, not the other way around.

Load-Bearing Assumptions on Which the Space Force Will Depend

Before considering some of the key assumptions about operations, training, acquisition, headquarters, and career management functions, it is useful to state five top-level load-bearing assumptions on which the entire Space Force concept rests. We summarize these assumptions in Table 7.1.

The first load-bearing assumption is that the United States needs a new organization to accelerate development of space doctrine, capabilities, and expertise to outpace future threats because the current approach is incapable of producing attractive alternatives to the status quo.

Although this assumption is reasonable in light of SPD-4 and the legislative language in the FY 2020 NDAA, its durability is not assured.[11] Preferences in the administration and Congress can shift with changes in leadership. The current Congress and administration have proceeded to establish the Space Force by legislation, but even as this Congress gives the go-ahead, there remains the possibility that a new Congress or administration could reverse course prior to the achievement of the FOC. Once the FOC is reached, the undoing of the decision to proceed becomes significantly more difficult administratively and politically.

The second load-bearing assumption is that a new organization (as opposed to other organizational or procedural measures that would chart a new course) will be an improvement over the status quo in terms of improving performance (however mea-

Table 7.1
Load-Bearing Assumptions Underlying the Sustainability of the Space Force

No.	Assumption	Plausible Time Frame
1	The United States needs a separate Space Force[a]	Achieved
2	The Space Force will improve effectiveness and efficiency over the status quo	Within the next 3 to 15 years
3	The Space Force's workforce will be sustainable	Within the next 5 to 15 years
4	The Space Force will gain sufficient independence	Within the next 5 to 15 years
5	The Space Force will develop a sense of identity	Within the next 5 to 15 years

NOTE: These time frames are not stated explicitly in official documents but instead represent our judgment of reasonable expectations.
[a] Pub. L. 116-92, 2019.

---

[11] White House, 2019; Pub. L. 116-92, 2019.

sured), minimizing duplication of effort, eliminating bureaucratic inefficiencies, and improving accountability. A new Space Force will almost certainly increase duplication of effort in some functions, most notably in headquarters and support activities. Bureaucratic inefficiencies are to be expected, particularly in the early years, as a consequence of having the Air Force and the Space Force where there had been only one service. Depending on the precision of initial cost estimates for the establishment of the Space Force, the ultimate costs and inefficiencies could be more than Congress had initially considered as reasonable (or not) and could lead Congress to slow (or speed up) the pace of spending toward the FOC. However, as noted in Chapter Three, the costs of duplication and inefficiency, at least for headquarters functions, are likely to be relatively modest, compared with the benefits of higher levels of control, mission focus, and ultimately warfighting performance that may result from the Space Force.

As one example: The strengths and weaknesses of this assumption will play out in the realm of PPBES processes. Current guidance dictates that the Air Force will continue to control PPBES processes in the Space Force, but how this is implemented in practice will matter. In principle, the Space Force will continue to use the same Air Force procedures and accounting systems to implement its PPBES activities. In practice, Air Force guidance will be imposed at the secretary level. Care will need to be taken to ensure that these procedures do not dictate substantive outcomes for the Space Force beyond under secretary guidance. Otherwise, there may be little reason to expect a different outcome from current arrangements.

A third load-bearing assumption is that the Space Force can achieve sufficient critical mass in its workforce to create viable pathways for career field development and advancement through its leadership ranks. Achieving sustainability in the Space Force workforce over time is by no means assured without deliberate steps taken to secure sufficient personnel in the key career fields. Detailed workforce analysis, planning, and monitoring would help to ensure that the most-critical career fields both survive and thrive.

A fourth load-bearing assumption is that the Space Force will achieve sufficient independence from the Air Force in its headquarters-level decisionmaking to distinguish itself in areas of concept development, force design, technological innovation, acquisitions tempo, and resourcing. Independence translates to the level of control in decisionmaking, often referred to as decision rights, vested in and exercised by the Space Force. In the discussion of headquarters activities noted in Chapter Three, senior leaders of the Space Force would likely have the authority to set parameters for and direction of development of concepts and doctrine. They also would want to consider exercising direct control over whatever direct reporting unit or field operating activity the Space Force creates to work out the details of new concepts and doctrine.

To ensure independence and a sense of identity within the Space Force, consideration could be given to transitioning resources and activities for development of concepts and doctrine to the direct control of the civilian secretary and the CSO. As another example in the realm of requirements, current guidance states that the Space Force will operate as an equal partner on the Joint Staff and represent DoD's primary space interests in JROC and JCIDS. These two roles (and associated authorities) will be central to the Space Force's role as an independent source of capability in a contested space environment, to translate combatant command priorities into engineering requirements, and to then advocate for space-related engineering requirements.

At the same time, failure or equivocation on the part of the Air Force and DoD about the Space Force's independence, perhaps most overtly expressed in terms of its ability to conduct legislative and public affairs activities, would undermine a central motivation for establishing the Space Force in the first place. Indeed, the effectiveness of the Space Force on the legislative front would be obvious: It would either win or lose in the House and Senate Armed Services Committees in advocating for its preferences. There may be degrees of winning and losing, but the ability of the Space Force to communicate its needs independent of the Air Force would become apparent in the Space Force's first congressional budget and appropriations cycle. If the Space Force is perceived to look and act too much like the current arrangement for space activities (no matter how effective these current arrangements may be), expectations for the Space Force as a "change agent" may be undermined.

Finally, a fifth load-bearing assumption relates to the ability of the Space Force to establish its own unique sense of identity among the military services that can help build cohesion and stability in its workforce and enable the Space Force to effectively conduct its missions. Building a sense of identity in any new organization carved out of existing organizations is difficult to do quickly. Rather, the process will likely happen organically through many individual decisions by Space Force leaders along the path from the IOC to the FOC, and beyond the FOC. The fate of this assumption is closely tied to the third assumption regarding sustainability of career fields, with each connected to the proposition that the Space Force will provide a professionally rewarding and meaningful work experience.

In the following sections, we consider the signposts and shaping and hedging actions that could mitigate the Space Force's vulnerability to failure of any one of these load-bearing assumptions.

Signposts

Several signposts could indicate whether initial decisions about consolidation of organizational units under the banner of the Space Force were working as anticipated. In cases in which signposts indicated that consolidation was going smoothly, actions could be taken to accelerate progress toward the FOC. In other cases, signposts could point to expected shortfalls in performance (where early assumptions have proven to be weak or invalid). In these cases, DoD or the Space Force could pursue actions to remove, shape, or hedge against these contingencies to forestall their occurrence or minimize their impact.

In deciding which operational components should be in or out of the Space Force, we applied four performance elements: the strategic objectives of effectiveness and efficiency and critical organizational attributes of independence and a sense of identity, first introduced in Chapter One and applied in Chapters Three and Four. However, making "clean" decisions based on these objectives and attributes carries judgments about the significance of unintended secondary effects of the transfer. For example, consolidation could degrade the effectiveness of joint force capabilities or efficiency in the use of resources *or* raise workforce sustainability concerns. Secondary effects that could complicate the transfer might occur if

- the transferred unit were also supporting other missions that were not assigned to the Space Force
- the transferred unit required service-unique training or expertise in addition to space expertise
- the transferred unit shared infrastructure with units supporting nonspace missions
- the transferred unit were operating in a location that was austere and required support that the Space Force would not typically provide
- the career field size were small, such that its sustainability in the Space Force could be challenging.

Experience over time will reveal whether the operational challenges that these factors present can be managed effectively and efficiently under the proposed consolidation plan. Procedures may be in place to grant some exceptions, but too many exceptions granted could eventually undermine the Space Force's effectiveness. Planners would need monitoring data and analytical tools to recognize whether their present decisions about how to consolidate operations would be effective and efficient over the longer term. In the face of changing concepts of operations, emerging threats and new technologies, they might consider keeping a watchful eye on whether their decisions may inadvertently increase mission risk elsewhere among joint forces and, at the same time, being mindful of impacts on both independence and identity.

In light of these potential secondary effects, the continued validity of the assumption of effectiveness and efficiency could be determined through periodic monitoring and reassessment. To know whether individual transfer and consolidation decisions were working as intended, the Space Force might wish to consider monitoring its own performance, as well as changes in the performance of the "originating" organizations from which operational units were transferred. Initiating an evaluation process prior to consolidation could ensure that critical baseline mission-specific performance metrics were gathered systematically. These "pre–Space Force" metrics would help planners make adjustments in the period from the IOC to the FOC, as well as beyond the FOC, as some negative impacts on originating organizations may take longer than others to emerge.

Signposts could include changes in budget requests of originating organizations as they seek to backfill staff and capabilities or requests for supplemental training beyond the capacities of the Space Force. President's budget requests and actual appropriations are other possible signposts. The Office of Management and Budget could push back on the Space Force's budget requests, or Congress could consistently under-appropriate resources for the Space Force. Other signposts could be an accumulating list of requests for waivers, exceptions, or delays requested to transfer some units, indicating that support for the Space Force may be eroding from within DoD.

In operational terms, signposts could be linked to mission effectiveness and indicate degradation in response times or resilience of systems to planned or unplanned disruptions. Signposts need not be limited to performance resulting from actions by the Space Force. In fact, the Space Force might consider signposts associated with each of the load-bearing assumptions to provide an early warning of when those assumptions begin to break down. One obvious signpost could come in the form of congressional action that could have an effect of fundamentally altering the direction of consolidation set in the IOC. Other signposts could parallel existing DoD metrics related to timeliness and quality of the acquisition process. Failure to shorten current timelines for acquisition of space technologies could prove to be a make-or-break issue for the Space Force. Another signpost could relate to the sustainability of space-critical career fields.

Shaping Actions

Depending on the signposts, the Space Force may choose to revisit some of its consolidation decisions, not only those that were implemented but also those that were delayed or rejected when the IOC was finalized. If this path were pursued, procedures could be developed to reverse or significantly reshape decisions that showed early signs of weakness or whose downside risks appeared, in practice, to outweigh the benefits initially considered in the IOC. Other procedures could be developed to help process lessons observed from the initial consolidation and then be applied to revised criteria for further consolidation to enable the Space Force to grow in ways consistent with its mission.

The idea behind a shaping action is to get ahead of emerging vulnerabilities and pursue a different course of action that would enable the Space Force to work around the problem or eliminate its impact entirely before real harm is done to organizational performance. An example of a shaping action could be a more active presence on Capitol Hill to help educate members of Congress and their staffs on the priorities of the

Space Force, the analytical basis for the Space Force's force structure decisions, and the need for resources to substantiate the priorities.

Another shaping action could relate to maintaining the third load-bearing assumption, of workforce sustainability. Ideally, the Space Force would be capable of managing its acquisition and other career fields to ensure that its personnel could provide the capabilities needed for full functionality. However, the small size of the Space Force may require special arrangements between the Space Force and the rest of the Air Force to train these personnel with skills relevant to operations in a contested space environment and the processes they staff and give future leaders the diversity and depth of experience they require to be leaders in the Space Force. Other shaping actions could include pursuit of authorization to raise staffing levels, provision of more-extensive training, and requests for additional congressional authorities to ensure that each of the Space Force's load-bearing assumptions remains firm.

Hedging Actions

DoD might not be able to shape future administration and congressional deliberations on whether to proceed with the Space Force as currently conceived, but it has available several hedging actions to advance the objectives of a new approach within its existing organizations. For example, with a strong monitoring and evaluation process in place, the Space Force could reinforce the premise of the IOC and continue to prioritize its consolidation activities, doing the "easy" moves first and then implementing the more challenging moves later, as it gains more capability to manage change, particularly regarding training and career field management. This could include measures by the Space Force to make increased use of Air Force headquarters and support services in the early years to enable it to focus instead on developing proficiency and credibility in its consolidated operations. These types of actions could help cushion the impacts of budget reductions or changes in leadership in the early years.

The discussion at the end of each major section of Chapter Three presented multiple examples of hedging actions that could help ease the transition to the Space Force and avoid a serious breakdown of functionality. With respect to career field management, hedging actions are distinguished by their impact on the health of the career pyramid, career paths, and the sufficiency of leadership opportunities. Each potential action could address issues that might arise in the transition from the IOC to the FOC, but they would need to be implemented quickly enough to forestall concerning trends in career field health that may appear in the signposts.

Potential Implications

We identified five load-bearing assumptions on which the success of the Space Force depends. Using the language of adaptive planning, we presented signposts (metrics) that the Space Force could choose to monitor and use to trigger shaping or hedging actions that could mitigate vulnerabilities to the IOC and, ultimately, the FOC. We provided examples of shaping and hedging actions regarding budgeting, sharing of services, and career field management that could be taken to mitigate some of the vulnerabilities that could emerge under threat of budget reductions or other significant changes in the internal and external environment as the Space Force matures from the IOC to the FOC and beyond.

Change in the Space Force's internal and external operating environments from what now appears to be the case is inevitable. The only question is what these changes will look like and when they may appear. To increase the likelihood of the survival and sustainability of the Space Force, it may wish to consider equipping itself with the tools to proceed in its implementation in an adaptive and agile manner. This could entail monitoring, analysis, and planning capabilities to anticipate and implement actions that could help mitigate, forestall, or avoid altogether the vulnerabilities to its performance that could undoubtedly arise.

As the examples in the chapter indicate, preserving the validity of the assumptions that the Space Force will achieve independence and identity is inherently in tension with the assumption that the Space Force will perform more effectively and efficiently than current arrangements. Space Force leaders will need to strike a balance. The inefficiencies will primarily affect activities within the headquarters of the Air Force and the Space Force, accounting for only a small fraction of the total resources used to create and sustain space-related capabilities. However, given the centrality of independence and identity to the character of the Space Force, inefficiencies may need to be tolerated to allow the Space Force to emerge and perform as envisioned.

CHAPTER EIGHT

Conclusion

By establishing the Space Force, the Trump administration is responding to growing threats to U.S. space operations with an organizational solution. The expectation is that having a separate armed service for space will create a champion within the U.S. armed forces to advocate for and develop new capabilities that can outpace current and projected threats. The RAND team has worked with the SFPTF to conduct analyses on how to best organize the Space Force under the guidance provided by the SPD-4, the legislative proposal, and the strategic overview.[1] To that end, we identified which forces in DoD should be consolidated within the Space Force. We also assessed whether the workforce within the Space Force would be sustainable.

We complemented these analyses with an examination of other organizations that DoD created to gain insights into potential challenges that the Space Force will face as it stands up and grows into its role. We drew on these insights in our assessment of which activities and organizations should transfer to the Space Force and in developing recommendations for the Space Force. We summarize our findings and recommendations below.

Space Force Activities and Organization

Chapter Four provides a detailed description of which space activities and organizations should transfer to the Space Force. These assessments were made based on organizational attributes of effectiveness, efficiency, independence, and identity.[2] Here, we broadly characterize the activities and organizations that should compose the Space Force and outstanding issues that need to be resolved (indicated in italics).

- **Operations and training:** The majority of the space operational units in DoD and the training units that support these operational units should be moved to

---

[1] White House, 2019; DoD, 2019b; DoD, 2019a.

[2] Chapter Three provides a detailed definition of these performance elements.

the Space Force. *Some space control units may reside in their current services until the Space Force's space control missions are further clarified.*

- **Acquisition:** The majority of space acquisition organizations (including SDA) should be consolidated in the Space Force. *Decisions about transferring MDA's space acquisition activities require more in-depth analyses.*
- **Headquarters:** Headquarters, Space Force, should include key functions that are essential to its independence and identity—operational concepts and doctrine development; requirements development and advocacy; planning, programming, budgeting, and execution; and legislative liaison and public affairs. *However, current policy guidance discourages transfer of some of these functions.*
- **Support:** The Space Force should leverage the Air Force for generic support services (for example, base operations and security, nonoperational logistics and communications, and professional services, such as medical, dental, legal, and contracting services). *The Space Force and the Air Force need to negotiate agreements on how to adapt standard Air Force approaches to Space Force priorities and how to fund such adaptation.*

Workforce Sustainability Challenges for the Space Force

To support these transfers, the Space Force would need personnel with skills in space operations, space intelligence, space acquisition, and other STEM disciplines relevant to space operations and acquisition. However, as discussed in detail in Chapter Five, our initial workforce sustainability assessment revealed major challenges:

- **Risks in sustaining small career fields:** Given the relatively small numbers in the Space Force, it will be challenging to sustain career fields as stand-alone or, in other words, to have uniformed Space Force personnel spend their full careers in the Space Force. This would be true even for the Space Operations career field (13S) as it is currently structured. To be sustainable, the 13S career field will need to redesign its billet structure to expand the O6 and command levels. A small but viable Space Force autonomous Intelligence career field appears to be sustainable within Space Force alone. However, analysis suggests that many Space Force career fields will need to share personnel with the Air Force. The Air Force's experience with small career fields suggests that they carry a higher risk, requiring the Space Force to plan and manage them carefully.
- **Shortfalls in general officer throughput:** Given the planned force numbers within the Space Force, and the estimated numbers of general officers that will be required, our quantitative analysis demonstrated that the Space Force's ability to generate sufficient general officers will be problematic. The Space Force should be able to generate, at most, about one-half of its general officer corps within the

foreseeable future. The structural issues with the 13S career field, the primary operators of the Space Force, would further exacerbate the problem, unless these issues are addressed. The bottom line is that the Space Force will likely need to draw about one-half of its general officers, in the foreseeable future, from the Air Force or other services.

Challenges and Opportunities for the Space Force

To be successful, the Space Force must seek to improve the effectiveness of the U.S. armed forces. Given the emerging threats to U.S. space-based assets, this is an important task.[3] From our analysis of other organizations, we found that, although other armed services often help enable joint force effectiveness, the Army, Navy, Air Force, and Marine Corps are all capable of directly employing violence in support of U.S. national security objectives as service components under a joint force commander. If the Space Force is limited to an enabling role, it could have difficulty demonstrating its effectiveness. In addition, space's current role as a force enabler could generate resistance from other services and agencies to turning over their current space activities, on which they depend, to the Space Force.

The Space Force will also face pressure to demonstrate that it is using resources efficiently. Creating a new service will inevitably result in some level of duplication, and a relatively small Space Force will not be able to enjoy many economies of scale. The Space Force will be able to consolidate some military space-related activities under one organization, but many military space-related functions will exist elsewhere. The Space Force will have a strong incentive to grow in terms of both end strength and funding, which could fuel charges of inefficiency. In addition, the small size of the Space Force relative to other services will result in it having less bureaucratic and political leverage in the defense community, potentially leaving the service vulnerable to second-tier status and possible reincorporation.

The Space Force will need to focus on developing and protecting its independence. Part of the motivation behind creating a separate Space Force is to create an organization focused primarily on developing U.S. space capabilities. SOCOM's ability to control some of its funds provides an example that the Space Force could usefully emulate. The fledgling Space Force will undoubtedly face considerable challenges as it competes with well-established services and other established organizations within DoD, but the Space Force will need to find some level of success to maintain a distinct place among the armed services.

---

[3] As defined in Chapter Three, *effectiveness* is "the level of military capability that the joint force can achieve within a fixed level of resources."

Last, the Space Force will have to create its own sense of identity. The identities of military organizations can change over time, driven by changes in the nature of armed conflict, popular support, and other factors. The Space Force will need to develop innovative CONOPs and warfighting doctrine to establish itself in the national consciousness as an essential military service with a distinctive mission. The Marine Corps can serve as a good model here, as it has adapted to changes in warfare and has developed a distinctive identity within the Department of the Navy. The lack of an existing coherent theory of space warfighting could hamper the Space Force's efforts. To firmly establish its identity, the Space Force will need to develop this body of theory, CONOPs, and doctrine to create an image of itself to audiences outside the armed forces, particularly in Congress and among the broader public.

Recommendations

We identified several areas that the Space Force should address to increase its likelihood of developing into a successful organization.

1. The Space Force Should Focus on Defining and Clarifying Its Space Warfighting Missions

Space control will be a core mission for the Space Force, as it will be at the heart of space warfighting.[4] In particular, it will be critical that the Space Force bring an independent, unique perspective on how space control will protect U.S. interests and improve the joint force's operational effectiveness in a more comprehensive way than how it is done today. However, the space control mission may remain fragmented in DoD because the services have different doctrine and CONOPs for how space control capabilities should be employed. Given this lack of common vision of space control doctrine and CONOPs, debates about transfers, command and control relationships, and resources related to space control missions will continue and potentially degrade the Space Force's ability to establish its identity and independence and hamper its ability to demonstrate the value of creating a separate service. We recommend that the Space Force make the following activities a high priority:

- **Develop and promulgate a coherent space warfighting theory** with which to teach Space Force members and foster the development of a distinctive identity and culture. To facilitate identity development, and acculturation, the Space

---

[4]   The *DoD Dictionary of Military and Associated Terms* defines *space control* as "[o]perations to ensure freedom of action in space for the United States and its allies and deny an adversary freedom of action in space." See Office of the Chairman of the Joint Chiefs of Staff, 2019, p. 198. Also see Joint Publication 3-14 (Joint Chiefs of Staff, 2018a).

Force will need to have as much control as possible over the conduct of its accession training, technical training, and professional military education.

- **Develop weapons and supporting systems to bring substance to theory** to make the Space Force a true warfighting service. Such capabilities need not be touted as independent war-winning weapons, as early airpower theory proclaimed. The U.S. military fights wars as a joint force—no one military service can be decisive acting alone. However, to justify its continued independence, the Space Force will need to bring important warfighting capabilities to the joint table that the other services cannot provide or, at least, have not provided to date.

2. The Space Force Must Control Its Own Resources and Public Relations

The Space Force will need control of its resources to select, advocate for, and acquire the weapons and systems needed to fulfill its warfighting missions. As the Air Force case study illustrates, if space warfighting activities reside within an armed service with a broader warfighting mission, that armed service may be reluctant to cut its other warfighting activities to provide resources for space warfighting activities. That armed service might not support programs for space systems that do not directly support their own operations. If the Space Force is to be truly independent, it will need independent resources, managed within a separate armed service.

The Space Force will also need to reach out to the American public to educate citizens on the nature and severity of space threats confronting the nation, the need for an independent Space Force to meet these threats, and how the Space Force will accomplish that mission. To survive over time, the Space Force will need a strong public constituency. However, current policy guidance states that the secretariat of the Air Force will continue to administer all planning, programming, and budgeting; public affairs; and legislative liaison activities for the Air Force and the Space Force.[5] Therefore, scale economies associated with these processes favor department-wide management over the creation of separate Space Force processes, placing the Space Force at a disadvantage. We recommend that the Space Force pursue the following:

- **Advocate for a separate total obligation authority for the Space Force.** The secretary of the Air Force should allocate a total obligation authority target each year that the Space Force will then work within to build its program and budget. Space Force programming and budgeting should be segmented from other Air Force programming and budgeting so that competition for obligation authority is applied only at the highest level within DAF. The Space Force should retain independent authority to advocate for its program and budget to the White House and Congress. The Space Force should also retain authority to move funds among its activities to the extent that Congress allows such flexibility. Such flexibility

---

5   DoD, 2019a.

should not extend to moving funds between activities in the Space Force and those elsewhere in the Air Force.

- **Advocate for the creation of distinct Space Force offices for legislative liaison and public affairs within the secretariat.** The secretary of the Air Force should give the Space Force authority to advocate for its own positions independent of these efforts for the rest of DAF.

3. The Space Force Should Find Creative Ways to Manage Its Career Fields and Develop Senior Officers

The small end strength will present many challenges in manning the Space Force and sustaining its career fields. Several essential career fields might not be large enough to stand alone in the Space Force, and thus the Air Force and the Space Force should collaborate on development of "space tracks" for those career fields that are not large enough to be organic to the Space Force but that require substantive space knowledge and skills. Additionally, challenges may present themselves in funding and resourcing billets (as highlighted in the NRO and AFGSC case studies) and in building a robust Space Force culture. Some reliance on civilian and contractor manpower may be helpful.

We recommend the Space Force pursue the following lines of effort to develop creative ways to manage its workforce:

- **The Space Force and the Air Force should find creative solutions to manage career fields and develop senior leaders.** The Space Force and the Air Force will need to build sustainable career field management and pathways to promotion within the Space Force. In this partnership, it will be beneficial for the Space Force to maintain some substantial level of independence in personnel, manpower, education, and training. The Air Force and the Space Force should collaborate on development of "space tracks" for those career fields that are not large enough to be organic to the Space Force but require substantive space knowledge and skills. The Space Force will not be able to function, sustain its career fields, and perform well using standard Air Force policies designed for much larger forces; it will need creative and flexible solutions.
- **The Space Force must work with Congress to meet its general officer requirements.** The Space Force must address the structure of the 13S career field, Space Operations, to ensure this core AFSC is sustainable and has sufficient O6 leadership and command positions. Additionally, as the Space Force becomes fully operational, the Space Force must determine how its general officer requirements will be manned, and, for the foreseeable future, it must work with Congress to draw from other services the approximately 50 percent of general officers that it will not be able to generate organically.

4. The Space Force Should Work Closely with Other Services and Organizations to Define and Manage New Seams and Relationships

Although consolidating space activities under one new service will reduce duplication and seams in certain areas, it will also create new seams and gaps in others. Managing these seams and new relationships will be essential to ensure effective integration of space with the other military operations they support on the land, in the air, and in the oceans. MDA has experienced similar challenges. As our MDA case study in Appendix A shows, separation from MDA and the operators it supports has contributed to interdepartmental friction and perhaps less warfighting capability. Additionally, the Space Force will need to rely on the Air Force for many generic services and infrastructure that are informed by Air Force policies. To ensure effectiveness and efficiency, the Space Force will need a mechanism to tailor these generic services to meet its needs.

We recommend that the Space Force and the other services take the following actions:

- **Retain appropriate organic space expertise within the Air Force.** With transfer of space-related expertise and forces to the Space Force, the Air Force will still need some of its own space expertise to ensure integration of Air Force and Space Force operations. The Air Force must sustain its own space expertise to support Air Force space needs.
- **Establish liaisons and strengthen their roles.** The Army, the Navy, and the Marine Corps should establish liaisons and new relationships with the Space Force to help Space Force leaders understand how to support their unique service requirements most effectively as they seek access to space-related expertise and services.
- **Increase space representation at combatant commands.** The Space Force should increase its representation at combatant commands to provide space support and its independent perspectives.
- **Craft formal agreements to specify and tailor services that the Air Force will provide to the Space Force.** Memoranda of agreement should specify what levels of service the Air Force will provide to the Space Force, how the Air Force will adapt these services to the Space Force's needs, who will pay for these services, and, in particular, who will pay for any adaptation of them.

5. The Space Force Should Establish a Formal Monitoring and Evaluation Process to Adapt to Changes in Organizational Priorities and External Factors

The plans that the SFPTF and the Space Force will develop to establish the service and grow into its role will likely evolve because of many uncertainties about how the threat environment will change; how the Space Force will be resourced and operated within the current and future DAF and DoD; how Space Force staff will be recruited, developed, managed, and retained; and how rapidly space-focused CONOPs and supporting technologies could be developed and produced. The Space Force will need to establish planning, monitoring, and analytic capabilities to anticipate and implement actions that can help mitigate, forestall, or avoid altogether the vulnerabilities to its performance that will inevitably arise. We recommend that the Space Force do the following:

- **Adopt an adaptive approach to planning.** The Space Force should establish an approach to planning that aims for robustness across a range of futures rather than optimality for a single (hoped-for) future, as well as the ability to adapt with agility to the inevitability of changes to initial planning assumptions.
- **Implement a formal monitoring and evaluation process as soon as possible to track "before" and "after" performance metrics.** Central to adaptive planning is the ability to monitor the internal and external environments for changes in planning assumptions. Space Force planners should develop a small set of well-chosen metrics related to effectiveness, efficiency, independence, and identity; establish baseline values of these metrics before implementation proceeds; and then monitor these metrics on at least a quarterly basis to determine the validity of initial assumptions about Space Force performance through the IOC and beyond.
- **Establish decision rules for acting, when necessary, based on information coming from the monitoring and evaluation process.** Using the information from the monitoring and evaluation process, Space Force senior leaders should determine when changes in the levels of the metrics will trigger shaping, hedging, or other adaptive actions to maintain the viability of the Space Force. In addition to identifying these decision points, senior leadership will need to ensure that changes consistent with the adaptive actions were made in the programming and budgeting processes.

Concluding Thoughts

Creating a service for space offers an opportunity to increase U.S. military capability, but undertaking an initiative of this significance also brings with it a range of challenges. To overcome these challenges the Space Force will need additional resources and flexibilities in relevant policies and processes in DAF and DoD. Since establishing a Space Force is a strategic decision for the United States, the new service should receive the resources and tools it needs to establish itself as an independent service and achieve its full potential to improve the joint force.

Case Studies: Other Examples of Organizational Change

This appendix contains a more detailed look at other military organizations. We start with the Army Air Corps and its transition to the Air Force. As the Air Corps successfully evolved into an independent department and service, the history of these two organizations is rife with implications for how the Space Force might evolve. Many readers of this report in the Air Force will be familiar with this history. We also consider the Marine Corps, which many view as an analogue to the Space Force because of the Marine Corps' status as a service within the Department of the Navy. CYBERCOM shows that to develop capabilities in another domain (in this case, cyber), DoD has decided to establish a combatant command instead of a service. The Title 10 responsibilities of SOCOM make it an interesting case to compare against the Space Force.

The small expected size of the Space Force, along with its role collecting intelligence, led us to include an examination of NRO, and the Space Force's role in enabling other missions makes MDA another interesting case for our purposes. Last, we consider Global Strike Command, which was created as a major command within the Air Force to address critical policy failures but has taken on a more lasting role within the service.

In each case, we consider the justification for establishing the organization, provide a brief history of how the organization evolved, and consider what implications the organization might have for the Space Force. We delve into more-historical detail for the Army Air Corps and the Air Force because of their heightened relevance for and shared history with the Space Force.

For a summary view of some of the implications we drew from our comparative work, see Table A.1.

From the Air Corps to the Air Force

When considering how to organize an independent Space Force, some might turn to the creation of the Air Force for inspiration and guidance. Indeed, a number of parallels suggest that the analogy is apropos. However, there are also differences between the two examples, and it is instructive to consider them and their implications for the

Table A.1
Comparative Implications for the Space Force

Organization	Year Founded	Impetus	Type of Organization	Key Implication
Air Force Global Strike Command	2009	Provide focus on strategic deterrence and global strike	Air Force major command	Need to secure resources and provide parity of rank with other organizations
CYBERCOM	2008	Unify cyber capabilities	Combatant command	Difficulty of managing through competing organizational interests
MDA	2002	Develop, test, and field an integrated, layered, ballistic missile defense system	Specialized DoD organization	Division between capability development and operational needs
SOCOM	1987	Conduct special missions and train partner militaries	Combatant command	Independent budget to provide autonomy
NRO	1961	Unify ISR systems	Specialized DoD and intelligence community organization	Difficulty of relying on other organizations for personnel
Air Force	1947	Develop capabilities in air domain	Independent service	Importance of developing warfighting theory, as well as weapons and supporting systems
Air Corps	1926	Develop capabilities in air domain	Branch inside a department	Importance of demonstrating capability in distinct medium
Coast Guard	1790	Ensure maritime safety, security, and stewardship within U.S. territorial waters	Uniformed service	Independent authority and budget to protect organization and help build distinct identity
Marine Corps	1775	Conduct ship-to-ship warfare	Service inside a department	Importance of creating a distinct culture and self-advocacy

organization of the Space Force, which we do in this section. It examines the organizational evolution of U.S. military air operations from where they began in the Army Signal Corps, which first acquired aircraft in 1907, to the independent Air Force, created by the National Security Act of 1947.[1] It highlights the theoretical, doctrinal, and cultural development of airpower as the driving force behind this evolution and reviews the political and technological developments that made it possible. Finally, the

---

[1]　Pub. L. 80-253, 1947.

section weighs the similarities and differences between the movement that resulted in a separate Air Force and efforts to create a Space Force and considers the implications for Space Force organization.

The Roots of a Long and Contentious Struggle

When the Army first began acquiring aircraft in 1907, its leaders considered the flimsy wood and canvas contraptions too fragile to conduct combat missions. However, they did see potential in the range, speed, and visual perspective they offered for communications and reconnaissance, so the Army created an aeronautical division for the new machines in the Signal Corps. Renamed the Aviation Section in 1914, air units remained there throughout most of World War I, until a lack of preparedness for America's entry into that conflict prompted a reorganization. The Army Air Service was organized as a combat branch in 1918 and formalized by an act of Congress in 1920.[2]

World War I provided the initial catalyst for the development of airpower and the movement for an independent air force. Although first relegated to support roles, it did not take long for airmen to find combat applications for the unique capabilities that their new technology provided. Almost immediately, opposing reconnaissance aircrews began shooting at each other, first with pistols, then with rifles, and ultimately with mounted machine guns, giving birth to pursuit aviation. The ability to travel far and fast, flying beyond enemy lines and attacking from above, gave rise to air interdiction, close air support, and even rudimentary strategic bombing missions. By the end of the war, all of the combat missions flown by today's air forces had emerged in some form.[3]

The Driving Force Behind the Air Force Independence Movement: Airpower Theory, Doctrine, and Culture

World War I was an important impetus for the development of airpower in another respect: The bloody stalemate at the trenches in France was such a debacle that it inspired military theorists in several countries to search for alternatives to repeating it. Airpower thinkers believed that they had the answer. Although aircraft had lacked sufficient capability to make a significant difference in the outcome of this war, these thinkers saw beyond the constraints of contemporary technology to grasp the potential of how aircraft might affect the outcomes of future wars. Aircraft would have the ability to strike the heart of an enemy's country without having to defeat its armies first. And though theorists debated what the principal targets of aerial bombardment should be, they agreed that the most effective and efficient use of airpower would be in strik-

---

2   Robert Frank Futrell, *Ideas, Concepts, and Doctrine: Basic Thinking in the United States Air Force 1907–1960*, Maxwell Air Force Base, Ala.: Air University Press, 1989, pp. 16–17, 34–35.

3   Lee Kennett, *The First Air War: 1914–1918*, New York: The Free Press, 1991, pp. 41–62.

ing targets deep in the enemy's homeland, such as cities and industry, instead of using it to support ground forces in their fight against the enemy army.[4]

Over the course of the 1920s and 1930s, Army airmen refined their theories into a coherent doctrine and used it to acculturate a growing movement among their fellow aviators. In 1920, the Air Corps Tactical School (ACTS) opened at Langley Field, Virginia, where it operated until 1931, when it moved to Maxwell Field, Alabama.[5] The school functioned there until 1940, when operations were suspended in anticipation of World War II. The central thrust of the doctrine that ACTS developed was that airpower was a weapon with war-winning potential, one best employed as an independent instrument against a country's vulnerable interior. That being said, not all members of the ACTS faculty agreed that airpower should be used exclusively for strategic bombing—Claire Chennault, who would later command the famous "Flying Tigers" in China, championed the merits of pursuit aviation—but they all agreed on at least one fundamental tenet: To achieve maximum concentration and flexibility, the employment of airpower should be centrally controlled. And that tenet had an essential corollary: Because only an airman can appreciate the proper use of airpower, its control should be centralized under the command of an airman, not an Army field commander.[6]

During the 20 years it operated, ACTS indoctrinated the many airmen who passed through its doors in a catechism of airpower. Airmen who could not attend in residence were reached via newsletters and correspondence courses. Over time, assigned mostly to airfields away from the Army mainstream and inculcated in an approach to warfare that contradicted the Army's doctrine for ground operations, the aviators developed their own culture and sense of identity. They became an insurgent force within the Army.[7]

---

[4] For a summary of the various early airpower theories, see Forrest E. Morgan, "The Concept of Airpower: Its Emergence, Evolution, and Future," in Richard B. Hallion, Roger Cliff, and Phillip C. Saunders, eds., *The Chinese Air Force: Evolving Concepts, Roles, and Capabilities*, Washington, D.C.: National Defense University Press, 2012.

[5] The school was first named the Air Service Field Officers' School when it opened in 1920, then renamed the Air Service Tactical School in 1922 and the Air Corps Tactical School in 1926. We use the last one because that was the name it held the longest and the one by which it is most often referenced in historical sources.

[6] Robert E. Finney, *History of the Air Corps Tactical School, 1920–1940*, Washington, D.C.: Air Force History and Museums Program, United States Air Force, 1998, pp. 55–78; Thomas H. Greer, *The Development of Air Doctrine in the Army Air Arm: 1917–1941*, Washington, D.C.: Office of Air Force History, U.S. Air Force, 1985, pp. 26, 72.

[7] Peter R. Faber, "Interwar US Army Aviation and the Air Corps Tactical School: Incubators of American Airpower," in Phillip S. Meilinger, ed., *The Paths of Heaven: The Evolution of Airpower Theory*, Maxwell Air Force Base, Ala.: Air University Press, 1997, pp. 183–238; James P. Tate, *The Army and Its Air Corps: 1919–1941*, Maxwell Air Force Base, Ala.: Air University Press, 1998, pp. 3, 148–149.

Political Pressure and Technological Development

Army and War Department leaders did not readily yield to the rising insurgency among Army aviators, but they were increasingly boxed in by pressure from the American public and Congress. Airpower advocates benefited from the romantic aura that surrounded aviation and the public enthusiasm it engendered. The availability of cheap surplus military aircraft prompted aviators returning from World War I to buy their own planes and begin touring the country, putting on "barnstorming" shows for local citizens. Demonstrations and record-setting feats, such as the sinking of the captured German battleship *Ostfriesland* via aerial bombardment in 1921, the flight around the world by an Air Service team in 1924, and Charles Lindbergh's solo transatlantic flight in 1927, galvanized the public imagination. Meanwhile, several senior Army Air Service officers grew increasingly vocal in advocating for greater air force autonomy. Most prominently, Brigadier General William "Billy" Mitchell, a charismatic figure who had been a senior U.S. aviator in France during the war, argued stridently that the air services should be granted separation from all Army and Navy control. In his view, air warfare was unique, and only airmen, whom he saw as a "distinctive class of . . . aerial knights engaged in chivalrous combat," had the proper mind-set to lead it.[8] He touted this romantic image at every opportunity in books, speeches, magazine articles, and newspaper editorials, often criticizing Army and Navy leaders for their hidebound attitudes and for mismanaging the air assets under their control.[9]

Under growing public pressure, the War Department and several congressional committees held board meetings and hearings in the 1920s and 1930s on questions regarding how to best organize and manage military airpower. In 1926, after several boards produced conflicting recommendations on the issue of air service independence, the House Military Affairs Committee fashioned a compromise, renaming the Army's air arm the Air Corps and providing a five-year plan for its expansion and development.[10] The expansion was delayed for several years by bureaucratic wrangling and constrained by a lack of funding during the Great Depression.

Then, after a move to have the Air Corps deliver mail in 1934 resulted in numerous crashes and 12 fatalities, the public outcry prompted the War Department to order a board of inquiry to investigate the fiasco.[11] The board found that the Air Corps,

---

[8] Mark A. Clodfelter, "Molding Airpower Convictions: Development and Legacy of William Mitchell's Strategic Thought," in Phillip S. Meilinger, ed., *The Paths of Heaven: The Evolution of Airpower Theory*, Maxwell Air Force Base, Ala.: Air University Press, 1997, p. 90.

[9] Ultimately, his public defiance of military authority resulted in his court martial and resignation from the Army, but he continued to extol the virtues of airpower and the need for an independent air force in publications and speeches until his death, in 1936. Mitchell's most-lasting contributions are his books: William Mitchell, *Our Air Force: The Key to National Defense*, New York: Dutton, 1921; William Mitchell, *Winged Defense*, New York: P.G. Putnam's Sons, 1925; and William Mitchell, *Skyways*, Philadelphia: J.B. Lippincott Company, 1930.

[10] Tate, 1998, pp. 45–47.

[11] John T. Correll, "The Air Mail Fiasco," *Air Force Magazine*, March 2008.

starved of funds under Army leadership, was ill-trained and ill-equipped. Nevertheless, the War Department rejected calls for the creation of an independent air force and, instead, endorsed a compromise proposed by an earlier board, which recommended the establishment of the General Headquarters (GHQ) Air Force to assume control over the Air Corps' peacetime training and operations and wartime employment from Army corps-level ground commanders. The chief of the Air Corps, operating in parallel, would retain responsibility for supply, procurement of personnel and equipment, schools, and doctrine development. The GHQ Air Force stood up in March 1935.[12]

As these events unfolded, technology advanced at a brisk pace, giving substance to the theoretical propositions of airpower and adding steam to the independence movement. The wood-and-canvas biplanes of World War I were replaced with all-metal monoplanes by the early 1930s. The development of variable-pitch propellers in 1933, coupled with new airframe designs, resulted in a doubling of airspeeds and operating ranges, seemingly overnight. Subsequent development of supercharged engines, cabin oxygen and pressurization, very high frequency radios, rotating gun turrets, and the Norden bombsight made possible the development of long-range heavy bombers, such as the B-17 Flying Fortress, which was test flown in 1935 and introduced to the force in 1938.[13]

Yet Army, Navy, and War Department leaders resisted the Air Corps' acquisition of heavy bombers for a variety of reasons, including concern that the bombers would strengthen the movement for air force independence. Preferring to spend available funds on less expensive light bombers, planes that would be used to support ground forces, Army Chief of Staff Malin Craig canceled B-17 procurement after the prototype crashed in 1935. Using a legal loophole, Air Corps leaders obtained funding to buy a small number of the aircraft for test purposes, thereby keeping the program alive. But in 1938, the Joint Army and Navy Board, of which Craig was the senior member, informed the secretary of war that it could foresee no use for a long-range bomber in a future conflict.[14]

Craig worried that allowing the Air Corps to purchase large numbers of B-17s "would merely encourage its fascination with strategic bombing, which remained untested and of dubious value."[15] The Army General Staff and War Department pre-

---

[12] The 1934 board was chaired by former Secretary of War Newton Baker. It consisted of 11 members, only three of whom were active or former air officers. The earlier board, convened in 1933, was chaired by Army Brigadier General Hugh Drum. It had five members, four of whom were senior Army ground officers. The 5 Drum Board members were among the 11 on the Baker Board. See John F. Shiner, "The Coming of the GHQ Air Force, 1925–1935," in Bernard C. Nalty, ed., *Winged Shield, Winged Sword: A History of the United States Air Force*, Vol. 1, Washington, D.C.: Air Force History and Museums Program, U.S. Air Force, 1997a.

[13] Martin W. Bowman, *The USAAF Handbook: 1939–1945*, New York: Stackpole Books, 1997.

[14] Phillip S. Meilinger, "When the Fortress Went Down," *Air Force Magazine*, October 2004; Greer, 1985, p. 99.

[15] John F. Shiner, "The Heyday of the GHQ Air Force, 1935–1939," in Bernard C. Nalty, ed., *Winged Shield, Winged Sword: A History of the United States Air Force*, Vol. 1, Washington, D.C.: Air Force History and Museums

ferred to spend available funding to rearm the ground forces with basic articles, such as rifles and mortars, versus glamorous, expensive aircraft. The Navy objected to the Air Corps procuring long-range bombers for fear that they would be used beyond the 100-mile coastal defense demarcation line previously negotiated between the services.[16] Given these concerns and the Joint Army and Navy Board's position, in August 1938, the War Department halted a program to develop a new high-altitude, long-range bomber and directed the Air Corps to order only twin-engine light bombers in 1939 and 1940.[17] Because of these obstacles, although the Air Corps had requested a total of 206 B-17s between October 1935 and June 1939, only 14 heavy bombers were delivered by the time Germany invaded Poland in September 1939.[18]

Air Force

The Army, Navy, and War Department resistance to arming the Air Corps was out of step with developments unfolding in Europe and Asia, and those developments soon overtook them. Alarmed by German aggression in the Sudetenland and Hitler's ability to intimidate European leaders with his Luftwaffe, President Franklin Roosevelt called a secret meeting with select cabinet members and military leaders at the White House on November 14, 1938, to discuss how to meet the rising threat.[19] There, the President shocked his top Army and Navy leaders by openly arguing for the mass production of aircraft.

Roosevelt said that he wanted an air force of 20,000 planes and an aircraft industry capable of producing 24,000 planes per year, although he doubted that Congress would approve more than 10,000 planes.[20] It would ultimately take several years to approach such numbers—not to mention to develop the infrastructure and personnel to operate such a force—but the President had made it clear that he wanted the nation to arm itself for war, and a major thrust of that armament would be a massive air force. The War Department's block on heavy bomber development and procurement was removed.

By the time the Army Air Corps was reorganized as the U.S. Army Air Forces (USAAF), it had been granted autonomy from the Army in all but name. In October 1940, Chief of the Air Corps General Henry "Hap" Arnold proposed that the Air Corps and GHQ Air Forces be consolidated to unify the air arm under one com-

---

Program, U.S. Air Force, 1997b, p. 145.

[16] Greer, 1985, p. 91.

[17] Futrell, 1989, pp. 86–87.

[18] Greer, 1985, p. 101.

[19] Attendees included secretaries of the Navy, war, and the treasury, as well as Craig, Air Corps Chief General Henry "Hap" Arnold, and Chief Naval Officer Admiral Harold Stark.

[20] Futrell, 1989, p. 91; Shiner, 1997b, p. 155; Greer, 1985, p. 100.

mander in preparation for war in Europe. The Army General Staff opposed this move, continuing to argue that, in the event of war, the Air Corps would have no mission separate from support of the ground forces.[21]

However, the effectiveness that British and German air forces, both independent services under unified command, demonstrated in battles in 1940 and early 1941, prompted Congress to begin questioning the "hydra-headed" command structure imposed on the U.S. air service. To forestall binding legislation, Army Chief of Staff General George Marshall and newly appointed Assistant Secretary of War for Air Robert Lovett worked out a compromise in which the air arm would be unified as the USAAF, with Arnold as its chief of staff, a position co-equal with the other service chiefs. The reorganization took effect in June 1941.[22]

In the years that followed, the USAAF established itself as an effective warfighting force, both in independent operations and in joint campaigns supporting surface forces. Because Germany had driven allied forces off the European continent before the United States entered the war, the only weapon with which U.S. and British forces could strike the enemy directly before June 1944 was airpower. Ardent believers in the ACTS doctrine of high-altitude, daylight, precision bombing, USAAF bomber groups coordinated with their Royal Air Force counterparts in the Combined Bomber Offensive to bomb industrial targets in Germany and occupied countries around the clock—the USAAF striking by day, and the Royal Air Force striking by night.

Similarly, USAAF units did long-range strategic bombing of Japanese cities, first from distant bases in China, then from bases in the Mariana Islands, once those strategic locations were captured in 1944. As these campaigns were taking place, USAAF tactical air forces supported U.S. and allied Army operations in North Africa, Sicily, and Italy in 1943, Western Europe in 1944 and 1945, and Army and Navy operations in the Pacific throughout the war. Over the course of the war, the USAAF grew into a massive force, one far larger even than Roosevelt had envisioned in 1938. In 1944, it reached a peak of approximately 2.37 million officers and airmen and 80,000 aircraft.[23] And though it shrank substantially during the drawdown after the war, it still comprised 305,827 men and women and approximately 30,000 aircraft when formally granted its independence in 1947.[24]

---

[21] Futrell, 1989, p. 103.

[22] Bernard C. Nalty, "Reaction to the War in Europe," in Bernard C. Nalty, ed., *Winged Shield, Winged Sword: A History of the United States Air Force*, Vol. 1, Washington, D.C.: Air Force History and Museums Program, U.S. Air Force, 1997, pp. 180–181.

[23] Office of Statistical Control, U.S. Army Air Forces, *Army Air Forces Statistical Digest, World War II*, Washington, D.C., December 1945, Tables 3 and 83, pp. 15, 135.

[24] *Air Force Magazine*, "The Air Force in Facts & Figures," in special issue, "2018 USAF Almanac," Vol. 100, No. 6, June 2018, p. 39.

By then, the Air Force had firmly established itself as an independent service, not just in law but in American popular culture. The strategic bombing campaigns in Europe and Asia had been skillfully publicized in military newsreels and romanticized in Hollywood movies. The contemporary Air Force narrative was that strategic bombing had been instrumental in winning the war in Europe, and Japan's surrender after the bombings of Hiroshima and Nagasaki proved the ACTS doctrine that airpower could be an independent, war-winning instrument. Although both assertions have been questioned, they were accepted by the American public and Congress immediately following the war, solidifying the Air Force's destiny as an independent military service.[25]

Air Force and Space Force Independence Movements: Similarities and Differences

Arguments for creating a separate Space Force resemble those used to justify Air Force independence in several ways. Most obviously, air and space operations are conducted in different domains from those of their parent services. As a result, they employ different physical principles and have different capabilities and vulnerabilities. Army airmen were quick to recognize that operating in the third dimension gave aircraft the ability to cover great distances in any direction quickly, free of obstruction by surface terrain. This offered prospects of taking war to the heart of an enemy's society, striking vulnerable targets that armies could not reach before defeating surface defenses. Given these capabilities, some Army airmen considered ground commanders ignorant and inflexible in their insistence that aircraft be used exclusively in support of ground forces—airmen wanted separation for the freedom to fully exploit the potential their new technology offered.

Space operators have made similar arguments. Satellites are subject to orbital mechanics—they move according to the principles of Kepler, versus those of Bernoulli, which govern winged flight in the atmosphere—and space systems are highly complex networked technologies that are significantly different from most air systems. Consequently, space operators are sometimes frustrated when they hear airmen trying to apply airpower tenets to space or describe space capabilities in ways that do not quite fit.

Operations in both domains have conjured romantic images in popular culture of intrepid people "pushing the envelope," venturing into new frontiers, selflessly risking their lives for science and country. Operators in both domains remained in their parent services for decades before gaining independence. And justification for separation in both cases has been made, in part, based on perceptions of rising threat from abroad—like the airmen before them, space operators, so the argument goes, need the freedom

---

25 For works addressing questions about airpower as an independent, war-winning instrument, see Gian P. Gentile, *How Effective Is Strategic Bombing? Lessons Learned from World War II and Kosovo*, New York: New York University Press, 2001; and Robert A. Pape, *Bombing to Win: Air Power and Coercion in War*, Ithaca, N.Y.: Cornell University Press, 1996.

and the resources to develop and employ new capabilities and doctrines to defend the United States against those being developed by potential enemies.

All of that being said, there are notable differences between the air and the space independence movements. The first is that aircraft were weaponized and employed in war very early in their history. Space capabilities, conversely, have not been employed as weapons. Although some rudimentary weapon technologies have been developed for experimental purposes, operational space capabilities have been used entirely in support of terrestrial forces. More significantly, aviators developed a coherent body of theory for employing air weapons in war, whereas space operators have not done so, despite numerous attempts.[26] Airpower theory was the driving force behind the Air Force independence movement. It was the intellectual basis and clarion call that unified Army aviators behind a shared vision, giving them a sense of separate identity and, in time, fostering a culture that differed from the Army mainstream. Rallying around this common vision, the Air Force independence movement was a bottom-up phenomenon, an insurgency within the Army.

The space community lacks this unifying force. Although many space operators see themselves as different from the Air Force mainstream, the stovepiped nature of space operations has tended to create multiple separate communities: space lift, satellite operations, space-based intelligence, missile warning, space surveillance, and so on. Without a unifying and widely embraced theory of space warfighting, it is questionable whether any sense of common identity and culture has emerged, or whether it can. Some space operators have opined that service independence is needed, but none has taken steps to move the community in that direction as the Army airmen did. The movement to create a Space Force is a top-down phenomenon.

Finally, although exploits in both domains have conjured romantic images in popular culture, there are important differences between them. Both have inspired a sense of awe in their scientific achievements; however, airpower has also been closely associated with heroism in war. Billy Mitchell's image of chivalrous "aerial knights," Claire Chennault's Flying Tigers, and the courageous bomber crews that, despite suffering enormous losses, persistently pounded Germany to rubble and set Japan ablaze in the cause of freedom stamped an image in the public psyche that was very different from that generated by such feats as the moon landing. Whereas the public largely credited airpower with winning World War II, when the Air Force achieved its independence, most Americans today associate space with exploration and scientific dis-

---

[26] For a few notable examples of efforts that have fallen short of developing coherent space warfighting theories, see David E. Lupton, *On Space Warfare: A Space Power Doctrine*, Maxwell Air Force Base, Ala.: Air University Press, 1998; James E. Oberg, *Space Power Theory*, Colorado Springs, Colo.: U.S. Air Force Academy, 1999; Charles D. Lutes and Peter L. Hays, eds., *Toward a Theory of Spacepower: Selected Essays*, Washington, D.C.: National Defense University Press, 2011. Conversely, for an example of space warfighting theory that is logically coherent but questionable in terms of political and technological feasibility, see Everett C. Dolman, *Astropolitik: Classical Geopolitics in the Space Age*, New York: Routledge, 2001.

covery, and some citizens in the United States and abroad recoil from proposals to put weapons in space, taking war to a supposedly unspoiled, peaceful domain.[27]

Implications of the Air Force Independence Movement for Space Force Organization

Despite the ways in which efforts to create a Space Force differ from the movement that led to Air Force independence, the risks associated with these differences can be mitigated. However, if the Space Force is to be successful as a warfighting service, able to stand on its own without being reabsorbed into its parent services, its leaders should try to replicate what made the Air Force successful to the extent possible. Several implications emerge from the Air Force independence example:

- The Space Force would need to develop and promulgate a coherent warfighting theory with which to indoctrinate its members and foster the development of a distinctive identity and culture. To facilitate indoctrination, identity development, and acculturation, it would need to have as much control as possible over the conduct of its accession training, technical training, and professional military education.
- The Space Force would need weapons and supporting systems to bring substance to its theory, making it a true warfighting service. Such capabilities need not be touted as independent war-winning weapons, as early airpower theory proclaimed. The U.S. military fights wars as a joint force—no one military service can be decisive acting alone. However, to justify its continued independence, the Space Force would need to bring important warfighting capabilities to the joint table that the other services cannot provide or, at least, have not provided to date.
- The Space Force would need control of its resources in order to select, advocate for, and acquire the weapons and systems needed to fulfill its warfighting missions. As the Air Force case illustrates, the Space Force's parent services would be unlikely to provide the resources needed for space weapon systems at the expense of cutting resources from their own programs. Nor would they be likely to support programs for space systems that did not directly support their own operations. If the Space Force is to be truly independent, it will need independent resources.
- The Space Force would need to reach out to the American public to educate citizens on the nature and severity of space threats confronting the nation, the need for an independent Space Force to meet these threats, and how the Space Force would accomplish that mission. To survive over time, the Space Force would need a strong public constituency.

---

[27] For instance, see Melissa de Zwart, "It's Not Clear Where Trump's 'Space Force' Fits Within International Agreement on Peaceful Use of Space," *Conversation*, June 19, 2019; and Chelsea Gohd, "Peace in Space: World Leaders Discuss Peaceful Uses of the Cosmos at UN Event," Space.com, September 22, 2018.

Marine Corps

Justification for Creation

The original impetus for the creation of the Marine Corps differs significantly from later justifications for the service's existence as an independent force. When the Marine Corps first formed in 1798 as a force within the Navy, the Marine Corps mission was to provide ship security and serve as the primary boarding element in naval combat.[28] When advances in naval artillery made boarding infeasible, the Marine Corps redefined itself as the primary force responsible for the capture and defense of advanced resupply bases for the Navy. In the modern era, the Marine Corps has again redefined itself as

> for service with the fleet in the seizure or defense of advanced naval bases and for the conduct of such land operations as may be essential to the prosecution of a naval campaign. In addition, the Marine Corps shall provide detachments and organizations for service on armed vessels of the Navy, shall provide security detachments for the protection of naval property at naval stations and bases, and shall perform such other duties as the President may direct. However, these additional duties may not detract from or interfere with the operations for which the Marine Corps is primarily organized. . . . The Marine Corps shall develop, in coordination with the Army and the Air Force, those phases of amphibious operations that pertain to the tactics, technique, and equipment used by landing forces.[29]

History

The Navy was formed in the image of its British and French counterparts.[30] Limits on the range, lethality, and rate of fire of gunnery meant that naval combat took place at ranges at which musket fire and boarding could be decisive.[31] As a result, a corps of Marines specialized in musketry and ship-to-ship fighting were seen as a necessity.[32] The Continental Marines was formed in 1775 and disbanded in 1783 after the signing of the Treaty of Paris. In 1798, the force was reestablished as a military service by an act of Congress that labeled the force the U.S. Marine Corps; the service was seen as an infantry force specialized in naval combat that would operate as individual detachments on naval vessels.[33] Marine Corps, Headquarters, handled the recruitment,

---

[28] Merrill L. Bartlett and Jack Sweetman, *Leatherneck: An Illustrated History of the U.S. Marine Corps*, Annapolis, Md.: Naval Institute Press, 2008, p. 34.

[29] 10 U.S.C. §5063.

[30] Bartlett and Sweetman, 2008, pp. 3–4.

[31] Bartlett and Sweetman, 2008, p. 4.

[32] Bartlett and Sweetman, 2008, p. 33.

[33] Kenneth W. Condit, John H. Johnstone, and Ella W. Nargele, *A Brief History of Headquarters Marine Corps Staff Organization*, Washington, D.C.: Historical Division, Headquarters, Marine Corps, 1971, pp. 2–3.

training, outfitting, and administration of new Marines, while remaining duties were managed by either the local Marine Corps commander or the Navy or Army officer that commander operated under.[34]

The 20th century introduced international and technological changes that affected both the mission and doctrine of the corps. The United States' victory in the Spanish-American War expanded U.S. overseas holdings, necessitating forward deployed resupply bases for the Navy's steam-powered fleet. Gunnery improvements extended the range of naval employment that made boarding unviable.[35] In an effort to maintain unity of command, the Navy turned to the Marine Corps to serve as its primary force to seize and defend advance resupply bases through amphibious operations.[36] The Marine Corps, in turn, became an expeditionary force with more-independent command that would allow its forces to operate far from shore. The Marine Corps expanded in size during both world wars, which required an expanded Headquarters, Marine Corps, to handle planning and administration.[37] This period also marked the effort to build a Marine Corps mythos within the public and the political mind.[38]

In the aftermath of World War II, the Army proposed a unification plan that, among other things, limited the Marine Corps to a regiment of 60,000 without organic air support.[39] Suspecting that the plan would significantly reduce the Marine Corps' ability to be a fighting force, the commandant harnessed the political support built through the efforts of the Publicity Department to directly petition Congress.[40] These efforts proved successful, as the National Security Act of 1947 made the Marine Corps a Title 10 entity.

Implications
Staff Structure Capable of Growth
The Marine Corps has experimented with different staff structures throughout its 221-year history. The Marine Corps' transition from the original three-department model toward a modified-general staff was driven by the inability of Headquarters, Marine Corps, to plan for and execute the rapid mobilization and expansion required by World

---

[34] Condit, Johnstone, and Nargele, 1971, p. 7.

[35] Bartlett and Sweetman, 2008, p. 149.

[36] Condit, Johnstone, and Nargele, 1971, p. 7.

[37] Condit, Johnstone, and Nargele, 1971, pp. 9–11.

[38] Aaron B. O'Connell, *Underdogs: The Making of the Modern Marine Corps*, Cambridge, Mass.: Harvard University Press, 2012, pp. 1–23.

[39] Bartlett and Sweetman, 2008, p. 303; Gordon W. Keiser, *The US Marine Corps and Defense Unification 1944–47: The Politics of Survival*, Washington, D.C.: National Defense University Press, 1982, p. 50.

[40] O'Connell, 2012, pp. 115–130.

War I.[41] During the interwar period the Headquarters, Marine Corps, realigned its planning section to match the function-centric General Staff Model, which was proved effective when the United States entered World War II.

The lesson for the future Space Force is this: Any organizational structure must possess enough planning capability to execute rapid growth and advocate its mission from a position of expertise.

Partnership and Competition

The Navy and the Marine Corps have a complex relationship within the Department of the Navy that entails partnership and competition. Partnership includes sharing the amphibious warfare operation, with the Navy procuring and manning major platforms, such as large-deck ships, while the Marine Corps provides such assets as infantry and vehicles for deployment in and from those ships. Partnership also includes the Navy providing the Marine Corps medical staff, chaplains, sealift, and education for certain career fields, while the Marine Corps furnishes ground troops specialized in seizing and holding littoral zones. A major area of competition between the two services is over budget top line, whereby each service's share of the Department of the Navy's funding total (typically 85 percent for Navy and 15 percent for Marine Corps) is fiercely guarded and, in "gray" areas, contested. The Department of the Navy's use of its own funds to develop, procure, operate, and maintain Marine Corps combat aircraft is an example of both partnership and intense competition. Heretofore unique to the Department of the Navy, the Navy secretary and under secretary adjudicate disagreements between the two services, and the financial management and budget organization within the Navy secretariat integrates, reprices, and finalizes two service program objective memorandum submissions.

The relationship between the Space Force and the Air Force within DAF will also likely involve partnership and competition. The partnership is rooted in space's long-standing enabling role for Air Force combat operations. The Air Force will continue to provide support for space-related bases and, likely, other areas not yet finalized. Competition is probable as the Space Force and Air Force vie for funding and could be especially vigorous if the Space Force seeks to expand in size or capability to achieve warfighting goals. Establishing set ground rules on funding allocations would limit bureaucratic conflict but could also reduce flexibility of DAF leadership in resourcing top priorities. The secretary of the Air Force and the secretary's financial management staff will face new adjudication and integration challenges.

Service Culture and Self-Advocacy

Coming out of World War I, the Marine Corps realized that it was small, dependent, and unpopular in both the public and the political eyes.[42] Over the interwar period,

---

[41] Condit, Johnstone, and Nargele, 1971, p. 9.

[42] O'Connell, 2012, p. i.

the Marine Corps built a warfighting-centric culture and narrative with which it leveraged heroism in World War II to create a support network both inside and outside Congress. When the President and the Army attempted to drastically limit the Marine Corps in 1947, the commandant was able to successfully leverage this support network to have the Marine Corps granted Title 10 status.[43]

Space Force planners should seriously consider how the newly organized force will build its own advocacy network and culture. A strong public affairs department able to engage with the media to shape narratives would be critical. Further, the new organization would need to instill in its members a culture built on the belief that whatever their mission is, only they have the capability to do it well. Admittedly, the Marine Corps has the advantage of having a predefined warfighting function and an operational history it can draw from. However, a culture of being on the technological and doctrinal cutting edge, such as seen at major tech firms, such as Tesla or SpaceX, might be leveraged instead.

Cyber Command

Justification for Creation

The core principle driving the creation of CYBERCOM was the need for single, unified cyber enterprise with a joint cyber workforce. In 1997, a no-notice military exercise, named Eligible Receiver 97, highlighted the inability of the U.S. government to recognize and respond to new and emerging cyber threats.[44] Over the years, as cyberspace matured, corresponding doctrine and organizations evolved in response. In 2004, the Joint Chiefs of Staff recognized cyberspace, with its abundance of malicious state and nonstate actors, as a unique warfighting domain—on par with the traditional warfighting domains of land, sea, and air.[45] The persistent and prevalent nature of cyber-based threats, ranging from cybercrimes to cyberattacks on defense networks, further highlighted the need for a well-trained, experienced cyber workforce. High-profile cyberattacks, such as the commercial hacking of defense contractors,[46] as well as the hacking of the Office of Personnel Management,[47] brought national attention to the importance of cyber as a domain.

Eventually, the necessity for a cyber workforce within DoD was codified into doctrine and national security strategy. For instance, *Achieve and Maintain Cyber-*

---

[43] O'Connell, 2012, pp. 268–280.

[44] William S. Cohen, "The 20-Year Climb to an Elevated CyberCom," *FCW*, October 9, 2017.

[45] CYBERCOM, "U.S. Cyber Command History," webpage, undated-b.

[46] Ryan Pickrell, "Chinese Hackers Are Reportedly Stealing Loads of US Navy Secrets, and the Navy Is Scrambling to Stop It," *Business Insider*, December 14, 2018.

[47] Michael Adams, "Why the OPM Hack Is Far Worse Than You Imagine," *Lawfare*, March 11, 2016.

space Superiority: Command Vision for US Cyber Command, a foundational document for CYBERCOM, states that cyber threats "transcend geographic boundaries and are usually trans-regional in nature."[48] Additionally, the document argues, "Disruptive technologies will eventually accelerate our adversaries' ability to impose costs."[49] Similarly, the 2018 National Cyber Strategy emphasized that "[c]yberspace is an inseparable component of America's financial, social, government, and political life,"[50] while also highlighting cyber activities of near-peer adversaries, such as Russia and China, and the use of cyberspace by both criminals and terrorists.[51] As a result, the importance of the cyber domain to national security and the global nature of cyber threats underpinned the ascension of CYBERCOM to a unified command.

History

CYBERCOM developed and evolved through several initiatives and institutional reorganizations spanning nearly two decades. On December 1, 1998, the Joint Task Force–Computer Network Defense, a component of the Defense Information Systems Agency, was "DoD's first organization to have authority to oversee and direct operations on individual military service and DoD networks."[52] By 2000, the task force evolved into the Joint Task Force–Computer Network Operations. In the same year, all cyber operations were collected under the auspices of Space Command. However, upon the dissolution of Space Command in 2002, U.S. Strategic Command (STRATCOM) assumed responsibility for the computer network defense missions for DoD.[53]

Under STRATCOM, the computer network defense mission was divided. STRATCOM's three-star deputy commander for network planning and integration, who served as the NSA director, led offensive computer operations. Meanwhile, STRATCOM's deputy commander for network operations and defense, serving as the director of the Defense Information Systems Agency, oversaw defensive cyber operations.[54]

By 2004, the Joint Task Force–Computer Network Operations was converted into Joint Task Force–Global Network Operations. This year was marked by the elevation of the cyber domain as a unique warfighting domain by the Joint Chiefs of Staff

---

[48] CYBERCOM, *Achieve and Maintain Cyberspace Superiority: Command Vision for US Cyber Command*, Fort Meade, Md., April 2018, p. 3.

[49] CYBERCOM, 2018, p. 3.

[50] White House, *National Cyber Strategy of the United States of America*, Washington, D.C., September 2018, p. 1.

[51] White House, 2018, p. 2.

[52] CYBERCOM, undated-b.

[53] CYBERCOM, undated-b.

[54] CYBERCOM, undated-b.

in the 2004 National Military Strategy.[55] In 2005, STRATCOM established Joint Functional Component Command for Network Warfare, which would be led by the director of NSA. Meanwhile, the Defense Information Systems Agency retrained control of Joint Task Force–Global Network Operations defensive mission.[56]

Under the direction of Secretary of Defense Robert Gates, the possibility of CYBERCOM was first discussed in early 2008. On November 12, 2008, Gates established CYBERCOM as a new subunified command under STRATCOM. On June 23, 2009, he ordered the merger of Joint Functional Component Command for Network Warfare and Joint Task Force–Global Network Operations and the creation of CYBERCOM. The subsequent merger codified the dual-hat relationship between NSA and CYBERCOM. By 2010, the newly founded command achieved IOC.[57]

On August 18, 2017, on the recommendation of Secretary of Defense James Mattis, President Trump elevated CYBERCOM to be the tenth unified combatant command. With its accession, CYBERCOM is responsible for cyberspace operations across the joint force. On May 4, 2018, General Paul M. Nakasone became the commander of CYBERCOM and director of NSA and chief of the Central Security Service.[58] The change of leadership ceremony simultaneously marked the official ascension of CYBERCOM to full combatant command status.

Implications
The Difficulty of Recruiting, Maintaining, and Retaining a Skilled Workforce

One of the principal challenges for CYBERCOM has always been the recruitment, training, and retention of a skilled and experienced cyber workforce. It is possible that the technology-dependent Space Force would face similar difficulties. Presently, the cyber workforce is composed of a combination of uniformed service members, contractors, and civilian employees. Under its current authorities, CYBERCOM has the authority to generate requirements and standards for the joint cyber force.[59] Nevertheless, each individual service is still principally responsible for the manning, equipping, and training mandate of the cyber force. This is largely accomplished through each service's cyber components, which report both to their respective service chiefs and CYBERCOM.[60]

---

[55] Joint Chiefs of Staff, *The National Military Strategy of the United States of America*, Washington, D.C., 2004.

[56] CYBERCOM, undated-b.

[57] CYBERCOM, undated-b.

[58] NSA, "NSA Welcomes General Paul Nakasone as Agency Director," May 4, 2018.

[59] Government Accountability Office, *DoD Training: U.S. Cyber Command and Services Should Take Actions to Maintain a Trained Cyber Mission Force*, Washington, D.C., March 2019, p. 30.

[60] James Stavridis and David Weinstein, "Time for a U.S. Cyber Force," *Proceedings*, Vol. 140, January 2014.

This arrangement creates a natural tension of institutional priorities, as each service operates within limited budgets and resources. Unsurprisingly, the needs of the cyber component are often subservient to the more-influential and well-established communities within each of the services. For instance, as of 2016, the Air Force's enlisted 1B3X airmen (Cyber Warfare Operations) was manned at only 46 percent.[61]

A 2019 Government Accountability Office report found that, "[a]s of November 2018, many of the 133 CMF [cyber mission force] teams that initially reported achieving full operational capability no longer had the full complement of trained personnel and did not meet CYBERCOM's readiness standards."[62] The report attributes lack readiness in the CMF to two major factors. First, each military service failed to provide adequate depth in their respective training implementation plans for the CMF. Consequently, this risked creating duplicative efforts and the inability to meet operation requirements set by CYBERCOM.[63] Second, CYBERCOM has not established a master training task list for foundational training courses across the services. This hinders the establishment of equivalency standards across services and risks inconsistent training across the CMF.[64]

Internal challenges of maintaining a ready cyber workforce are also exacerbated by external market forces. As of September 26, 2018, the Pentagon was planning to hire 8,300 cyber specialists in the next year.[65] Cyber hiring goals are challenging, given that cyber and technical skills are highly sought after, in both the public and private sectors. Unsurprisingly, the public sector, such as the military, has struggled to compete with its private-sector counterparts. During the CyberCon 2018 conference, Gregg Kendrick, the executive director of the U.S. Marine Corps Forces Cyberspace Command, commented that the disparity in compensation between the Marine Corps and the private sector is stark—creating a significant challenge in recruiting sufficient talent.[66] General Richard Angle similarly commented, "This is a very high-demand space with a low density population to draw from."[67] The challenge of recruiting a highly technically skilled workforce is not limited to CYBERCOM. In recent years, NSA has also struggled to recruit and retain its workforce. The lack of competitive

---

[61] William E. Parker IV, *Cyber Workforce Retention*, Maxwell Air Force Base, Ala.: Air Force Research Institute, Air University, October 2016, p. 37.

[62] Government Accountability Office, 2019, p. i.

[63] Government Accountability Office, 2019, pp. 20–21.

[64] Government Accountability Office, 2019, pp. 25–26.

[65] Justin Lynch, "Inside the Pentagon's Struggle to Build a Cyber Force," *Fifth Domain*, October 18, 2018a.

[66] Justin Lynch, "Why Recruiting Cyberwarriors in the Military Is Harder Than Retaining Forces," *Fifth Domain*, November 1, 2018b.

[67] Lynch, 2018a.

pay and several national scandals involving NSA impropriety have reportedly created a talent drain on the agency.[68]

The growing rift between DoD and the private technological sector has further exacerbated the issue. In 2018, employees of Google successfully petitioned the company to drop Project Maven, a DoD artificial intelligence project, over moral and ethical objections.[69] Likewise, Microsoft employees have petitioned for Microsoft to drop its augmented reality contract with the Army.[70] If the rift between Silicon Valley–based companies and DoD continues to grow, the challenges of recruiting the necessary talent for high-skilled public-sector roles, such as CYBERCOM and NSA, will become only more arduous.

The Challenge of Service Equities

Prior to CYBERCOM's formation, each respective military service manned, equipped, and trained its own service cyber force and developed its own service-oriented network with little or no coordination across the joint force. CYBERCOM was established to combine and centralize the DoD cyber force from the disparate cyber forces scattered throughout the services. In essence, CYBERCOM was to serve as the joint integrator, provide an operational headquarters for cyber operations, and generate requirements and standards for the cyber force.

Some of the justification for the Space Force has been that it will serve to integrate space capabilities across DoD. At this point, however, there seem to be significant challenges to achieving this goal. The planned independence of NRO, along with Army and Navy reluctance to part with their space capabilities, could lead to continued inefficiency and confusion as the Space Force moves forward.

During the time CYBERCOM existed within STRATCOM, there was intense debate about whether the newly recognized cyber domain required a new stand-alone cyber force on par with other military services. Advocates of a separate cyber force argued that the centrality of cyberspace in modern conflicts and power projection demanded a unique service to meet operational demands. Furthermore, cyber force advocates argued that cyber equities would never be properly served within the other services, which prioritize their own respective warfighting domains.[71] Unsurprisingly,

---

[68] Ellen Nakashima and Aaron Gregg, "NSA Is Losing Talent over Low Pay, Flagging Morale and Reorganization," *Chicago Tribune*, January 2, 2018.

[69] Drew Harwell, "Google to Drop Pentagon AI Contract After Employee Objections to the 'Business of War,'" *Washington Post*, June 1, 2018.

[70] Paresh Dave, "Microsoft Workers Demand It Drop $480 Million U.S. Army Contract," Reuters, February 22, 2019.

[71] Stavridis and Weinstein, 2014.

the newly formed CYBERCOM "seemed the 'odd man out' in a command focused on nuclear deterrence and strategic strike."[72]

Meanwhile, critics of a separate cyber force argued that the cyber domain largely served as an enabling domain of the traditional warfighting domains of land, sea, and air.[73] Moreover, critics argued that cyber did not meet the same requirements that the Air Force met to justify its own service—particularly in terms of new doctrine of warfare, political support, and technological maturity.[74]

As of today, despite being elevated to a unified command, CYBERCOM remains in an enabling and supporting role of the other military services. However, the debate over the necessity of a stand-alone cyber force persists. In August 2018, James Stavridis wrote that the nation required a separate cyber force—"a unified command with a unique service culture."[75] Nevertheless, CYBERCOM has, in large part, been unable to truly acquire and centralize DoD's disparate cyber capabilities and forces. The military services have been unwilling to relinquish control of their respective cyber components. As of now, CYBERCOM has been unsuccessful in overcoming entrenched service equities in the broader cyber community.[76]

Competing Organizational Interests

Originally, the co-location and dual-hatted nature of the newly formed CYBERCOM and the more established NSA were designed to generate institutional efficiencies. The arrangement enabled CYBERCOM to build on the extensive physical infrastructure and institutional knowledge that already existed within NSA. The close relationship between the two organizations also allowed for greater coordination, faster decision-making, shared resources, and cost-saving measures.[77]

However, the competing nature of each respective organization proved problematic, highlighting the conflicting perspectives of Title 10 and Title 50 organizations.[78] The differing objectives of a military-centric CYBERCOM and an intelligence-centric NSA routinely led to mutually exclusive objectives—leading to internal tension and frustrations. As the more established and senior organization, NSA routinely overrode

---

[72] Thomas Spoehr and James DiPane, "Elevating Cyber Command: An Overdue Step Towards Enhancing Military Cyber Operations," Heritage Foundation, October 1, 2018.

[73] Mark Pomerleau, "Rogers: Cyber Doesn't Needs Its Own Military Branch," *Defense Systems*, January 21, 2016.

[74] Corey M. Ramsby and Panayotis A. Yannakogeogros, "A Reality Check on the Cyber Force," *Strategic Studies Quarterly*, Summer 2016.

[75] James Stavridis, "The U.S. Needs a Cyber Force More Than a Space Force," *Bloomberg*, August 14, 2018.

[76] Christopher Paul, Isaac R. Porche III, and Elliot Axelband, *The Other Quiet Professionals: Lessons for Future Cyber Forces from the Evolution of Special Forces*, Santa Monica, Calif.: RAND Corporation, RR-780-A, 2014, pp. 29–30.

[77] Government Accountability Office, *Defense Cybersecurity: DoD's Monitoring of Implementing Cyber Strategies Can Be Strengthened*, Washington, D.C., August 2017, p. 12.

[78] 50 U.S.C.

CYBERCOM operations in favor of intelligence-gathering efforts. In reference to the cyber campaign against the Islamic State, former Secretary of Defense Ash Carter characterized the dilemma by stating, "When CYBERCOM did produce something useful, the intelligence community tended to delay or try to prevent its use, claiming cyber operations would hinder intelligence collection."[79] A 2017 Government Accountability Office report also stated that the dual-hat nature of leadership led to "increased tension between NSA/CSS [Central Security Service] staff and CYBERCOM staff."[80]

CYBERCOM provides an illustrative example of the challenges of combining two institutions with distinctive cultures, incentives, and identities. The intertwined organizational structure between NSA and CYBERCOM provides benefits in efficiencies, especially in terms of resources and deconfliction.[81] Nevertheless, the power and influence of institutional identities and equities have continually fueled the debate about whether the two organizations should be permanently split. As of now, the two organizations remain joined, as decisionmakers have preferred to weigh tangible technical expertise over the intangibles of culture and institutional equities.

In the near future, CYBERCOM and NSA could potentially be separated into two distinct, independent organizations. The 2017 Government Accountability Office report highlighted that, "as CYBERCOM reaches maturity, the number of operations will likely grow, and the breadth, depth, and magnitude of the issues required to be managed by a single person leading both CYBERCOM and NSA/CSS could be overwhelming."[82] The recent elevation of CYBERCOM to full unified combatant command status only increases the possibility of a permanent separation, as the span of control becomes untenable for both organizations.

The fledgling Space Force could face a similar tension between Title 10 and Title 50 organizations as it sorts out its relationship with NRO and with other intelligence agencies that operate in space. Separate funding lines, different authority structures, and different constituencies would likely continue to create seams between the Space Force and other U.S. government entities operating in space.

---

[79] Ash Carter, *A Lasting Defeat: The Campaign Against ISIS*, Cambridge, Mass.: Belfer Center for Science and International Affairs, Harvard Kennedy School, October 2017.

[80] Government Accountability Office, 2017, p. 13.

[81] James DiPane, "Should Cyber Command and the NSA Have Separate Leadership? How to Decide," Heritage Foundation, May 2, 2019, pp. 5–6; Government Accountability Office, 2017, p. 11.

[82] Government Accountability Office, 2017, p. 12.

Special Operations Command

Justification for Creation

After a series of embarrassing and deadly operational failures, congressional leaders sought to reform the special operations community. The result was the creation of SOCOM in 1987. The functional command was mandated to ensure that the U.S. military possessed a credible and elite special operations capability. Given the specialized and joint nature of special operations forces (SOF) missions, unique and expansive authorities, such as command and control; training, organizing, and equipping; and special acquisitions, were centralized in SOCOM.

Dedicated special forces units have existed since World War II, but these forces were organic and separate elements within their respective services. However, following the Vietnam War, considerable mistrust and tension existed between the conventional military establishment and SOF. This period of decline and neglect is best reflected in the failure of Operation Eagle Claw and command and control issues during the invasion of Grenada.[83] Subsequent congressional pressures led to the creation of the Joint Special Operations Agency in 1984. Yet it is important to note that the agency possessed neither operational nor command authority over any service-retained SOF.[84] Through the 1987 NDAA, the modern incarnation of SOCOM would be established in April 1987.[85]

History

Over time, the missions, roles, and authorities of SOCOM have evolved in accordance with changing geopolitical and military demands. As a uniquely empowered unified combatant command, SOCOM centralized operational control of all SOF units, and SOCOM now fosters interoperability between services and boasts its own funding authority through the MFP-11, also known as the "SOF checkbook."[86] U.S. Code, Title 10, Section 167, mandated that a four-star commander would lead SOCOM, along with an assistant secretary of defense for special operations and low-intensity conflict to provide civilian oversight.[87] Some of SOCOM's responsibilities were considered "service-like" and unique among the other combatant commands, in that SOCOM managed its MFP-11 funding and was responsible for SOF-specific acquisition, following its designation with "extraordinary authority over SOF force structure, equipping,

---

[83] SOCOM, History and Research Office, *United States Special Operations Command History: 1987–2007*, Tampa, Fla., 2007, p. 5.

[84] SOCOM, History and Research Office, 2007, p. 5.

[85] Pub. L. 99-661, National Defense Authorization Act for Fiscal Year 1987, November 14, 1986.

[86] SOCOM, History and Research Office, 2007, p. 7.

[87] 10 U.S.C. § 167; SOCOM, History and Research Office, 2007, p. 7.

and resourcing."[88] Some in DoD were reluctant to accept the creation of SOCOM and its role among the military services, as evidenced by a speaker at SOCOM's activation ceremony saying that the command would need to "break down the wall that has . . . come between [SOF] and the other parts of the military . . . [and] educate the rest of the military."[89]

SOCOM remained flexible and changed over the years under the guidance of each new commander. Its activities later came to include counterproliferation, information operations, and counterterrorism, and its organizational structure shifted.[90] The global war on terror (GWOT) resulted in the greatest changes of SOCOM. The devastation of September 11 resulted in an immediate search for a solution to the terrorist network growing around the world. Some believed that SOCOM was the answer, as SOF provided immediate support to the geographic combatant commands after September 11, and Secretary of Defense Donald Rumsfeld wished for SOCOM to be the single lead in the GWOT. Yet across DoD, the roles and responsibilities of SOCOM relative to the geographic combatant commands and the theater special operation commands were heavily debated.

If SOCOM were to lead the GWOT effort, it was unclear whether it would continue its service-like functions of training and equipping SOF or whether these responsibilities would return to the services.[91] Other concerns were the loss of regional expertise from the geographic combatant commands, which were reluctant to give up their responsibilities. The Unified Command Plan ultimately addressed these and other concerns, but it was a contentious process among the geographic combatant commands, Joint Staff, and the Office of the Secretary of Defense to finalize it.[92] At the Combatant Commanders' Conference in 2005, General Bryan Brown, the commander of SOCOM at the time, suggested that SOCOM be the "synchronizing" authority among the combatant commands for efforts against terrorist networks. This meant that SOCOM "would have authority to compel the combatant commands to mesh their counterterrorist plans and operations with SOCOM's campaign," with SOCOM being the lead for GWOT plans and operations.[93] Although the commanders voted eight to one against the proposal, with the eight instead in favor of a plan in which SOCOM had a coordinating authority only, Rumsfeld and President George W. Bush decided that SOCOM would be the synchronizing authority for the GWOT and declared such a few months later.

---

88 SOCOM, History and Research Office, 2007, p. 13.

89 SOCOM, History and Research Office, 2007, p. 7.

90 SOCOM, History and Research Office, 2007, p. 12.

91 SOCOM, History and Research Office, 2007, p. 15.

92 SOCOM, History and Research Office, 2007, p. 15.

93 Quoted in SOCOM, History and Research Office, 2007, p. 16.

The GWOT and its ever-expanding global mandate elevated SOCOM's prestige and relevance as the employment of SOF forces exponentially increased—becoming the force of choice in the GWOT.[94] The other combatant commands then had to adapt their plans and operations in accordance with SOCOM's, and SOCOM maintained the authority to train, organize, and equip its SOF personnel. The structure of the combatant command shifted again to fit its new demands, as did its funding. Prior to 2001, SOCOM's annual budget was about $4 billion; for FY 2006, it had increased to $7.4 billion and continued to grow.[95] The 2006 Quadrennial Defense Review had such an emphasis on the role of special operations that SOCOM, essentially, became like a fifth service.[96] SOCOM also created a new component, a Marine Corps special forces component, to join its Air Force, Army, Navy, and joint components.[97] In 2008, SOCOM became responsible for security force assistance efforts, and in 2013, it received the responsibility of training, organizing, and equipping the theater special operation commands, for which the geographic combatant commands had previously been tasked.[98]

Implications
Funding Authorities and Unique Prerogatives
By design, SOCOM combines both the operational authorities of a typical unified command and select responsibilities of a stand-alone service. Like a service, SOCOM is responsible for organizing, training, and equipping roughly 71,000 SOF.[99] To execute its wide-ranging responsibilities, the command is endowed with unique prerogatives. The command's MPF-11, comprising its budget, and its expedited procurement process serve as the best examples. Through the MPF-11 and Overseas Contingency Operations funding, roughly totaling $10 billion, SOCOM wields tremendous autonomy and independence from the other parent service.[100] Combined with the command's use of indefinite delivery/indefinite quantity contracts, SOCOM is able to rapidly acquire equipment for its specialized needs.[101] As a result, SOCOM is able to be innovative, flexible, and highly specialized. With Title 10 responsibilities, the Space Force should be able to enjoy similarly tailored authorities and an ability to develop

---

[94] Zimmerman et al., 2019, pp. 126, 136.

[95] SOCOM, History and Research Office, 2007, p. 21.

[96] DoD, *Quadrennial Defense Review Report*, Washington, D.C., February 6, 2006; SOCOM, History and Research Office, 2007, pp. 21–22.

[97] SOCOM, History and Research Office, 2007, p. 22.

[98] Congressional Research Service, 2019.

[99] Zimmerman et al., 2019, p. 125.

[100] Zimmerman et al., 2019, p. 133.

[101] Paul, Porche, and Axelband, 2014, p. 52.

capabilities uniquely suited to the space environment. Freed from having to compete with Air Force priorities, the Space Force should be better positioned to develop space-appropriate forces and CONOPs.

Establishing a Joint yet Specific Institutional Identity

Reflecting its distinct role in the DoD ecosystem, SOCOM has cultivated and institutionalized a pervasive culture of jointness. Despite the persistence of service components within SOCOM, the command has built a common and inclusive mantra of interoperability and cooperation. This shared institutional culture, spanning different services, is a product of both history and constructed culture. For instance, the SOF community draws on a long history of unconventional warriors, marked with salient historical moments and acts of heroism.[102] Furthermore, according to a former special operations unit commander, ostracization of special force operators from their parent service has allowed them "to forsake their service culture in pursuit of excellence."[103] As a result, the SOF community and SOCOM benefit from a unifying culture as exemplified by the "Five SOF Truths."[104] Through this institutional identity, SOCOM acts as a persistent and powerful advocate for the needs of the SOF community—operationally, institutionally, and politically. In a similar fashion, separation from the Air Force could allow the Space Force the opportunity to create a unique identity focused on its mission instead of having to develop an identity in the midst of the Air Force.

Coast Guard

Justification for Creation

The Coast Guard traces its institutional origins to the immediate postrevolutionary period, when the survival of the young nation's nascent domestic industry was threatened by wartime debts and British imports. As part of a comprehensive economic proposal to generate income and protect domestic production through customs duties and tonnage taxes, Secretary of the Treasury Alexander Hamilton proposed to Congress a series of measures to improve maritime security, aid navigation, and extend the federal government's capacity to regulate the trade of goods. On July 31, 1790, Congress enacted An Act to Regulate the Collection of the Duties Imposed by Law on the Tonnage of Ships or Vessels, and on Goods, Wares, and Merchandise Imported into the United States, which established a "system of cutters" to enforce federal tariffs and cus-

---

[102] Paul, Porche, and Axelband, 2014, p. 44.

[103] Zimmerman et al., 2019, p. 129.

[104] Zimmerman et al., 2019, p. 133.

toms laws on the nation's sea trade.[105] Known alternately as the System of Cutters, the Revenue Source, and the Revenue-Marine until it was formally renamed the Revenue Cutter Service (RCS) in 1863, the new organization was placed under command of the secretary of the treasury and tasked with the single mission of assisting the collection of customs duties and tonnage taxes.[106]

History

Since its establishment, the Coast Guard's mandate has broadened to include 11 statutory missions encompassing both homeland security and non–homeland security duties. Although the initial motivations for the service's establishment were economic, Congress and the executive branch have justified the continued expansion of the Coast Guard's authorities on security, humanitarian, economic, and environmental grounds. As the United States' territory and international influence grew and the scale of trade and maritime traffic reaching U.S. shores increased, Congress and the executive branch looked to the Coast Guard as a flexible instrument that might fulfill regulatory, law enforcement, and defense functions. The Coast Guard's unusually broad mandate is the product of an incremental series of executive orders and federal legislation, reinforced by successive executive branch reorganizations, that consolidated the service's jurisdiction over activities related to or conducted on and through U.S. waterways and sea-lanes, ships and maritime vessels, coasts, ports, and other maritime facilities.

The contemporary Coast Guard is the product of a series of restructurings designed to centralize authority over maritime safety and security. First established as the RCS within the Department of the Treasury, the service was renamed the Coast Guard when it merged with the U.S. Life Saving Service in 1915. The Coast Guard later assimilated the Steamboat Inspection Service, Bureau of Navigation, and U.S. Lighthouse Service, streamlining the process of establishing and enforcing maritime regulations.[107] In 1967, the service was transferred to the newly formed Department of Transportation in fulfillment of President Lyndon Johnson's desire to bring all federal transportation authorities into a unified department. Despite efforts to maintain the Coast Guard's identity as a military service, the reorganization resulted in a downgrad-

---

[105] On August 4, 1790, Congress expanded the act with An Act to Provide More Effectually for the Collection of the Duties Imposed by Law on Goods, Wares and Merchandise Imported into the United States, and on the Tonnage of Ships or Vessels.

[106] U.S. Coast Guard, *Doctrine for the U.S. Coast Guard*, Washington, D.C., Coast Guard Publication 1, February 2014, pp. 28–29; U.S. Coast Guard, *U.S. Coast Guard Missions: A Historical Timeline*, Washington, D.C., June 25, 2017.

[107] William H. Thiesen, "The Coast Guard's World War II Crucible," *Naval History Magazine*, Vol. 30, No. 5, October 2016.

ing of the organization's defense duties and capabilities in favor of its civil regulation and law enforcement missions.[108]

With the transition to the Department of Homeland Security in early 2002, the modern Coast Guard has prioritized homeland defense as its core mission but continues to fulfill its traditional port security, search and rescue, and drug interdiction missions. Coast Guard personnel and assets protect foreign pipelines and international shipping lanes, contribute to humanitarian and environmental relief efforts, conduct counterpiracy missions in international waters, and provide support for combat operations.[109] In short, the Coast Guard serves to protect the United States from threats from the seas, on the seas, and to the seas.

Homeland Defense

Despite its initial intention to employ the newly formed cutter service for tariff and customs enforcement, Congress called on the RCS to defend the homeland when tensions with France intensified in 1799. As the United States' sole standing maritime force, the Revenue-Marine was assigned to protect the nation's coastlines and to defend its interests on the high seas, with expanded congressional authority to board, search and examine, and certify the manifest of all U.S.-bound ships sailing within four leagues of the United States.[110] These duties expanded to include combined naval and land operations during the later Seminole and U.S.-Mexican Wars.[111]

As the United States' international influence grew, the RCS's homeland defense mission strengthened in response to new foreign challenges. With the service's renaming in January 1915, President Woodrow Wilson designated the Coast Guard as a Title 10 authority, establishing the basis under which the Coast Guard was transferred temporarily from the Treasury to the Navy during World Wars I and II. During both wars, Coast Guard personnel and assets were authorized to escort U.S. and allied convoys, control shipping in major ports, and prevent sabotage at the nation's harbors.[112] Wartime pressures also incentivized the expansion of the Coast Guard's civil jurisdiction to inspect merchant vessels and documentation, oversee personnel licenses, and operate facilities with long-range aids to navigation.[113] The world wars, therefore,

---

[108] Coast Guard Aviation Association, "United States Coast Guard Transferred to the Department of Transportation," in *United States Coast Guard Aviation History*, Troy, Va., 2017.

[109] Lawrence J. Korb, Sean Duggan, and Laura Conley, *Building a U.S. Coast Guard for the 21st Century*, Washington, D.C.: Center for American Progress, June 2010.

[110] U.S. Coast Guard, 2017.

[111] U.S. Coast Guard, 2017.

[112] U.S. Coast Guard, 2017; Thiesen, 2016. Evoking Wilson's expansion of the Coast Guard's duties in times of war, President Franklin Roosevelt also ordered the Coast Guard, in 1940, to oversee the anchorage and movement of all vessels in U.S. waters and to defend U.S. vessels, harbors, and waterways; the Coast Guard was assigned authority over all vessels with high explosives and other dangerous cargoes.

[113] Thiesen, 2016.

established the precedent for greater cooperation between the Coast Guard and the Navy and laid the groundwork for the service's subsequent foreign deployments.

The Coast Guard's transfer to the newly formed Department of Homeland Security in February 2003 marked the final major shift in the organization's priorities. The reorganization elevated homeland security to the Coast Guard's primary mission and augmented the service's authorities to safeguard U.S. ports and waterways. The guard established new domestic mobile Maritime Safety and Security Teams to provide operational commanders with quick-response capabilities to protect commercial high-interest vessels, Navy high-value assets, and critical waterside infrastructure.[114] The Maritime Transportation Security Act of 2002 further strengthened the Coast Guard's authorities by designating Coast Guard Captains of the Port as the Federal Maritime Security Coordinators and codifying the service's status as the lead agency responsible for coordinating all maritime security planning and operations in U.S. ports and waterways.[115]

Drug Interdiction and Other Law Enforcement

The Coast Guard's law enforcement authorities swelled with the size of the U.S. economy and federal government over the 19th and 20th centuries. When Congress outlawed the importation of enslaved persons into the United States in 1807, the RCS assumed responsibility for enforcing the law on the sea.[116] These authorities were strengthened further after the War of 1812, when Congress augmented the RCS's duties to include the enforcement of quarantine and neutrality laws and the protection of merchant vessels from piracy. Even as resources shifted to the Navy, the passage of the Posse Comitatus Act in 1878 protected the RCS's jurisdiction by designating the RCS as the sole enforcer of federal law on the high seas and U.S. waters.[117] Thereafter, the RCS was called on to enforce federal law in newly acquired coastal territories, a process that required concurrent augmentations to its law enforcement authorities and capabilities.[118] This process of incremental growth continued beyond the period of continental expansion, as Congress continued to delineate specific enforcement responsibilities to the Coast Guard, including authorization to prevent the maritime

---

[114] U.S. Coast Guard, 2017.

[115] U.S. Coast Guard, 2014, p. 15; Pub. L. 107–295, Maritime Transportation Security Act of 2002, November 25, 2002.

[116] U.S. Coast Guard, 2017.

[117] U.S. Coast Guard, 2014, p. 11; U.S. Coast Guard, 2017; Pub. L. 45-263, Posse Comitatus Act of 1878, June 18, 1878.

[118] U.S. Coast Guard, 2017.

importation of illegal alcohol (1919), enforce the Alaska Game Law (1925) and Whaling Treaty Act (1932), and lead a variety of other antismuggling measures.[119]

The Coast Guard's maritime capabilities, antismuggling experience, and broad authority to make inquiries, examinations, inspections, searches, seizures, and arrests on the high seas prepared the service to assume a major role in countering the illicit trade of narcotics. The Coast Guard intensified its antinarcotics operations over the 1980s, establishing the National Narcotics Border Interdiction System, which began operations in 1983, and improving its coordination with national and international organizations to identify interdiction targets and coordinate joint actions.[120]

Migrant Interdiction and Humanitarian Relief

When the RCS merged with the U.S. Life-Saving Service in 1915, the resulting organization, renamed the Coast Guard, became the United States' sole maritime service dedicated to safety and security at sea. The Coast Guard assumed a variety of humanitarian duties and capabilities designed to aid in the evacuation of U.S. citizens, the provision of command-and-control and search-and-rescue capabilities, and the transportation and distribution of vital supplies.[121]

As the United States' primary maritime law enforcement agency, the Coast Guard also became the first responder and enforcer of U.S. immigration law at sea. This migrant interdiction mission was established in 1794, when Congress tasked the RCS with enforcing a new prohibition on U.S. citizens' participation in the transportation of enslaved persons, but expanded dramatically after 1808, when the passage of a complete prohibition on the importation of enslaved persons led Congress to expand the RCS's jurisdiction to the high seas.[122] It was not until the large-scale exodus of Cubans from the island in the 1960s, however, that the Coast Guard confronted the challenge of mass maritime migration that the guard improved coordination with other federal agencies and foreign countries to identify vessels carrying migrants, interdict and rescue vessels in danger, and secure maritime routes of entry into the United States, its territories, and its possessions. The Coast Guard's duties continued to expand with subsequent waves of maritime migration to include the arrest of alien smugglers, the facilitation of the parole for prosecution or repatriation of non-U.S. citizens found committing criminal acts at sea, the inspection of maritime vessels and facilities to

---

[119] Pub. L. 66-66, National Prohibition Act, October 28, 1919; Alaska Game Law of 1925, Ch. 75, 43 Stat. 739, January 12, 1925; Whaling Treaty Act of 1932, 49 Stat. 1246, March 31, 1932; and Timber Reserve Act of 1822, 3 Stat. L. 651, February 23, 1822. On the RCS's role in countering smuggling, see Andrew Wender Cohen, "Smuggling, Globalization, and America's Outward State, 1870–1909," *Journal of American History*, Vol. 97, No. 2, September 2010.

[120] Pub. L. 98-473, The National Narcotics Act of 1984, October 12, 1984.

[121] U.S. Coast Guard, 2014, pp. 51–55.

[122] Irving H. King, *The Coast Guard Under Sail: The U.S. Revenue Cutter Service, 1789–1865*, Annapolis, Md.: Naval Institute Press, 1989, chapters 1–2.

ensure workers' legal employment status, the temporary detention of undocumented immigrants, and the handling of all requests for asylum made while in Coast Guard custody.[123]

Traditional Missions: Marine Safety, Aid to Navigation and Ice Operations, Marine Resources, and Environmental Protection

In addition to its homeland defense and law enforcement functions, the Coast Guard fulfills a variety of safety and regulatory functions that date back to the service's origins in the late 18th century. By tasking the Department of the Treasury with responsibility for constructing and maintaining the nation's system of lighthouses in 1789, Congress established a precedent to task the service with a range of functions related to improving the safety of and ease of navigation through U.S. waterways and facilities.[124] Similarly, the passage of legislation authorizing the Treasury to regulate the steamboat industry after a series of deadly boiler explosions in 1852 laid the groundwork for the service's broad regulatory authorities and the establishment of a centralized maritime inspection service.[125] The Coast Guard's unique assets, in turn, prepared the service to assume additional responsibilities over systematic oceanography, maritime observation, and the navigation of icy waters that required the service to develop the capacity to operate jointly with state and international actors.[126]

Finally, the Coast Guard fulfills a variety of marine resource management and environmental protection and regulation functions that date to the late 19th century and include the protection of endangered species;[127] the enforcement of federal regulations related to pollution, maritime resources, and environmental protections;[128] and

---

[123] For a review of the Coast Guard's legal authorities, see Gary W. Palmer, "Guarding the Coast: Alien Migrant Interdiction Operations at Sea," *Connecticut Law Review*, Vol. 29, 1996, pp. 158–179; Azadeh Dastyari, *United States Migrant Interdiction and the Detention of Refugees in Guantánamo Bay*, New York: Cambridge University Press, 2015, chapter 3.

[124] U.S. Coast Guard, 2014, pp. 30–31, 36.

[125] Steamboat Act of 1852, 10 Stat. L., May 30, 1852. Despite the Coast Guard's rapid growth over the 19th century, it competed with several independent agencies tasked with targeted missions. In a wartime measure on February 28, 1942, the Roosevelt administration consolidated all the nation's maritime safety functions within the Coast Guard, a change that became permanent five years later. Pub. L. 219, An Act to Integrate Certain Personnel of the Former Bureau of Marine Inspection and Navigation and the Bureau of Customs into the Regular Coast Guard, to Establish the Permanent Commissioned Personnel Strength of the Coast Guard, and for Other Purposes, July 23, 1947; The Motorboat Act of 1940, Ch. 155, 54 Stat. 163, April 25, 1940; Pub. L. 85-911, Boating Act of 1958, September 2, 1958; Pub. L. 92-75, Federal Boat Safety Act of 1971, August 10, 1971.

[126] See, for instance, White House, "Interdepartmental Board on International Service of Ice Observation, Ice Patrol, and Ocean Derelict Destruction," Washington, D.C., Executive Order No. 2458, September 20, 1916.

[127] See, for instance, the Coast Guard's role in enforcing the 1925 Alaska Game Law and the 1932 Whaling Treaty Act (Alaska Game Law of 1925, Ch. 75, 43 Stat. 739, January 12, 1925; Whaling Treaty Act of 1932, 49 Stat. 1246, March 31, 1932).

[128] U.S. Coast Guard, 2014, pp. 51–54.

the collection of meteorological, oceanographic, and other scientific data, among other areas. As in the previous mission areas, the Coast Guard's marine resource and environmental protection duties grew steadily but incrementally in response to technological advances and political pressure to improve regulations after major safety incidents or disasters.[129]

Implications for the Space Force

The Coast Guard's history and organization provide insights for developing a future Space Force. First, the Coast Guard's incremental expansion over the course of its history illustrates the relationship between innovation and institutional growth. The Coast Guard's 11-mission mandate is the result of a series of targeted and crisis-driven legislative amendments to address emerging technologies, which established, often unintentionally, the legal precedent for subsequent and more-expansive authorities. The Coast Guard's experience, therefore, underscores the importance of maintaining enough institutional flexibility to react to evolving political, technological, and regulatory pressures.

Similarly, the Coast Guard's long history inoculated the organization against the crises of identity that might be expected to follow each reorganization. That the service can trace its origins to the early republic has offered its members a distinct institutional identify and lent the organization a degree of coherence that belies its unusual evolution. The Space Force may need to lean on its shared history with the Air Force until it is able to establish its own distinct culture.

As the smallest of the military services, the Coast Guard's unique organizational structure could be used as an example by the future Space Force. Because of its small size and unique mission set, the Coast Guard headquarters has combined many of the traditional Napoleonic staff structure under the deputy commandant for military support and the deputy commandant for operations, with many of the subdepartments lead by civilian Senior Executive Service members. This combined organization allows the Coast Guard to operate a leaner headquarters element while concentrating its officer corps in the regional commands.

Finally, the Coast Guard's Title 10 authorities allowed the service to maintain its homeland defense capabilities through multiple institutional reorganizations and despite its status under civilian and law enforcement agencies. The secure line of funding allowed Coast Guard leaders to resist bureaucratic pressure from their organizations and to maintain baseline assets and capabilities that could be expanded during wartime and after their transition to the Department of Homeland Security. The legal authority conferred by Title 10 also reinforced the service's self-identity as a military

---

[129] Public outrage after the 1989 Exxon Valdez oil spill, for instance, spurred Congress to augment the Coast Guard's emergency rapid-response capabilities and to increase its authority to regulate potential vulnerabilities within the oil industry, such as tanker construction and personnel licensing. U.S. Coast Guard, 2014, pp. 51–54.

service and provided a unifying narrative. The future Space Force would benefit from having its own funding stream.

Missile Defense Agency

Justification for Creation

In 1999, the United States faced multiple strategic problems. The Cold War had ended a little under a decade prior, but nuclear weapons remained a principal currency of the relationship between the United States and Russia.[130] And even as the former Soviet threat waned, Iran and North Korea were developing capabilities that threatened the security of U.S. allies, U.S. troops abroad, and even the U.S. homeland. One way to meet these challenges was to double down on missile defense capabilities. A layered ballistic missile defense system that could deny a limited missile attack against the U.S. homeland or U.S. allies would preserve U.S. strategic freedom of action against Iran and North Korea. Further, reorienting the U.S.-Russia relationship away from mutually assured destruction could lessen suspicions between the two former foes and allow for further reduction of nuclear arsenals.[131] However, U.S. missile defense efforts—despite robust funding—had failed to produce a field-ready ballistic missile defense system on the scale envisioned. These are the problems that spurred the creation of MDA.

National security policymakers saw mutually assured destruction, of which the Anti-Ballistic Missile Treaty was an important issue, as an outdated model for deterrence and security. Since the Soviet Union fell, and Russia was not considered an adversary, a new model of deterrence was needed. Policymakers wanted to make the U.S.-Russian relationship about more than nuclear weapons.[132] However, previous efforts to develop and field missile defense on the scale required to fundamentally change nuclear deterrence had failed. Despite $40 billion or more in investment, the United States had little to show for operational ballistic missile defense for theater, regional, or intercontinental threats.[133] President Ronald Reagan launched the Strategic Defense Initiative with similar intentions, but no comprehensive missile defense system appeared likely. The effort might have been stillborn given the enormity of the task and lackluster record of missile defense. But the administration enjoyed a wave of

---

[130] Condoleezza Rice, *Missile Defense Papers*, Washington, D.C.: White House, 2001.

[131] Rice, 2001.

[132] Rice, 2001.

[133] Committee on Armed Services, House of Representatives, *Declaration of Policy of the United States Concerning National Missile Defense Deployment*, Washington, D.C., 106-39 Part 1, 1999; Donald Rumsfeld, *Missile Defense Program Direction*, Washington, D.C.: Department of Defense, 2002.

bipartisan support spurred largely by the growing threat of rogue regimes developing ICBMs tipped with weapons of mass destruction.

In 1998, Secretary Rumsfeld led the Commission to Assess the Ballistic Missile Threat to the United States, later known as the Rumsfeld Commission. The commission found that the U.S. national security community undervalued the threat posed by rogue states developing ballistic missiles. Rather than consider ballistic missile development by adversaries as a matter of contraband, the commission recommended that the United States treat ballistic missiles as instruments of state power developed by states for strategic purposes.[134] Congress, when debating what became the 1999 National Missile Defense Act,[135] cited the findings of the commission. Ultimately, the committee voted 50 to 3 to declare that the policy of the United States was to deploy a national missile defense.[136]

The committee report emphasized and concurred that the threats posed by ballistic missiles and weapons of mass destruction were more advanced and more rapidly advancing than current intelligence suggested. Further, the committee concurred that the United States would not have sufficient warning to meet threats posed by ballistic missiles as they became apparent.[137]

Of particular concern was the increasing availability of ballistic missile information in the open domain, where rogue actors, such as North Korea and Iran, could access it and advance their own designs. Further, the committee named China and Russia as state actors that were aiding such nations in acquiring and fielding ballistic missile systems, another finding of the Rumsfeld Commission.[138]

The committee then named two current events driving the change in policy: launches of ballistic missiles by both North Korea and Iran.[139] Most concerning to the committee was North Korea's test of the Taepo Dong-1 ballistic missile, an event the intelligence community failed to anticipate. The missile's third stage theoretically gave it the capability to reach portions of the United States, such as Alaska. Concurrently, North Korea continued work on the Taepo Dong-2 ballistic missile, whose 10,000-kilometer range put most of the continental United States at risk.[140]

The problem with North Korea possessing an ICBM was that it incentivized North Korea to use it in a crisis or conflict. North Korea possessed—then and now—inferior conventional forces. The Kim regime was and is unlikely to survive a con-

---

134 Donald Rumsfeld, "Intelligence Side Letter to the Rumsfeld Commission Report," March 18, 1999.

135 Pub. L. 106-38, National Missile Defense Act of 1999, July 22, 1999.

136 Committee on Armed Services, House of Representatives, 1999, p. 5.

137 Committee on Armed Services, House of Representatives, 1999.

138 Committee on Armed Services, House of Representatives, 1999.

139 Committee on Armed Services, House of Representatives, 1999.

140 Committee on Armed Services, House of Representatives, 1999.

vectional conflict with South Korea or the United States. The addition of a nascent ICBM capability gave the regime two more choices to end a crisis or conflict on favorable terms. One, North Korea could quickly fire a portion of its ballistic missiles at the United States or its allies. In this scenario, North Korea ensures that it leverages its ICBM capability prior to potentially losing it in the opening blows of a conflict. Two, North Korea reserves ICBM strikes until the Kim regime feels sufficiently threatened. In either case, North Korea demonstrates the capability and willingness to inflict further punishment on U.S. and allied civilian populations.

Rather than accept defeat, North Korea could then hold hostage those populations it has yet to strike. For the United States and its allies, the strategic conundrum becomes choosing between risking further civilian casualties in their respective homelands or ending the conflict on Pyongyang's terms.[141] Prior to developing ICBM capabilities, the North Koreans had little means to inflict pain on the U.S. homeland or even allied homelands outside the Korean Peninsula.

This threat increases the potential costs of U.S. influence and action on the Korean Peninsula and potentially erodes deterrence against a North Korean who may see a viable path to coercive diplomacy. Another potential strategic issue is the decoupling of U.S. alliances. If Japan or South Korea feels that the United States is willing to press a crisis while it receives ballistic missile attacks but will relent once the U.S. homeland is attacked, it may be reticent to back the United States in a crisis in the first place and may even seek to pacify North Korea.

Ballistic missile proliferation increases risk to U.S. homeland, troops abroad, allies, and friends. This was most clear in the case of North Korea. Ballistic missile technology was proliferating and evolving, and the capabilities or rogue regimes, such as North Korea and Iran, seemed on track to overwhelm mitigating capabilities if current progress (or lack thereof) continued.

Rumsfeld converted the Ballistic Missile Defense Organization to MDA, seeking to elevate the stature of missile defense and streamline research, development, and operational fielding of a comprehensive national ballistic missile defense system.[142] He tasked MDA to manage, direct, and execute the development of the ballistic missile defense system and to achieve the following DoD priorities:

1. Defend the United States, deployed forces, allies, and friends from ballistic missile attacks of all ranges in all phases of flight.
2. Develop and deploy, as directed, a layered ballistic missile defense system.
3. Enable the fielding of elements of the ballistic missile defense system as soon as practicable.

---

[141] David A. Ochmanek and Lowell Schwartz, *The Challenge of Nuclear-Armed Regional Adversaries*, Santa Monica, Calif.: RAND Corporation, MG-671-AF, 2008.

[142] Rumsfeld, 2002.

4. Provide capability in blocks, improving the effectiveness of fielded capability by inserting new technologies as they become available.[143]

History

MDA is the latest formulation of the U.S. effort to develop—and now field—missile defense capabilities. Over time, the United States adjusted its approach to missile defense based on the emergence of threats, policymaker approaches to foreign policy, and technological advancements. Generally, policymakers have hewn to either limited missile defense or attempted to develop missile defense against even the most robust attack. The original charter of MDA has a mixed record. It did field world-class regional missile defense (THAAD), as well as dozens of Ground-Based Midcourse Defense systems, but has fallen well short of a comprehensive missile defense that fundamentally changes the U.S.-Russian deterrence relationship.[144]

Early U.S. missile defense efforts are characterized by a lack of centralized control and sporadic attention from policymakers. The U.S. military identified the threat posed by the Nazi German V-2 ballistic missile in the aftermath of World War II. Nazi Germany fired V-2s at the United States during World War II. The United Kingdom attempted to counter the V-2 by pairing early radar identification and tracking with intense antiaircraft fire. Overall, the approach failed to destroy incoming V-2s in sufficient numbers. The U.S. military recognized the potential threat posed by ballistic missiles and undertook multiple efforts across the military services to develop effective countermeasures.[145]

U.S. missile defense took on renewed importance when the Soviet Union demonstrated an effective SS-16 ICBM and successfully launched Sputnik over the course of three months in 1957.[146] The U.S. responded with a statement of policy to protect assets that directly contribute to the defense of the North American continent and the retaliatory capability based on the North American continent.[147] The Secretary of Defense tapped the Army to lead continued development of the anti-ICBM capabilities, most notably the Nike Zeus. Ultimately, the bulk of operationalized missile defense was passive: hardened silos, alert programs, and early warning.[148] The technol-

---

[143] DoD, *Missile Defense Agency (MDA)*, Washington, D.C., Department of Defense Directive 5134.09, 2009.

[144] MDA, "Elements: Ground-Based Midcourse Defense (GMD)," webpage, last updated October 3, 2019a.

[145] John Dabrowski, *Missile Defense: The First Seventy Years*, Fort Belvoir, Va.: Missile Defense Agency, 2013.

[146] Dabrowski, 2013.

[147] National Security Council, *U.S. Policy on Continental Defense*, Washington, D.C., NSC 5802, 1958.

[148] Dabrowski, 2013.

ogy for antimissile missiles was too immature and could not hope to deploy in sufficient numbers to meet the growing Soviet ICBM threat.[149]

President Richard Nixon subsequently focused ballistic missile defense on defending U.S. Minuteman ICBM silos against Soviet attack. Nixon used the Safeguard system—originally planned to provide terminal coverage of the national capital region and ICBM silos—as bargaining chips in the first Strategic Arms Limitation Talks (SALT I) and the Anti-Ballistic Missile Treaty talks. Consequently, the United States shuttered the sole operational Safeguard site shortly after it started operations.[150]

President Reagan launched the first major consolidation and restructuring of the U.S. missile defense: the Strategic Defense Initiative. His stated aim was to move the U.S.–Soviet Union deterrence relationship from deterrence by punishment to deterrence by denial—that is, using missile defense to deny the Soviet Union the ability to destroy the United States in a nuclear conflict.[151] Reagan launched the Strategic Defense Initiative (otherwise known as "Star Wars"), which emphasized research and development of missile defense. After a year of study, multiple missile defense programs were consolidated underneath the Strategic Defense Initiative Organization. President George H. W. Bush ordered a review of the Strategic Defense Initiative program after the fall of the Berlin Wall and subsequently reoriented missile defense once again to defeating limited strikes. He announced the Global Protection Against Limited Strikes system in his 1991 State of the Union Address. President Bill Clinton continued the trend away from defending against large-scale attacks, instead focusing on compliance with the Anti-Ballistic Missile Treaty and theater missile defense. In 1993, he broke up the Global Protection Against Limited Strikes and renamed the Strategic Defense Initiative Organization the Ballistic Missile Defense Organization.[152]

As discussed, the George W. Bush administration substantially changed the course of U.S. missile defense. The administration formed MDA to rapidly meet the threat of rogue states developing ballistic missiles, shift away from a nuclear deterrence–centric relationship with Russia, and overcome bureaucratic impediments to fielding missile defense.

Implications for the Space Force

The MDA model—specifically, an agency outside the JROC with its own unique acquisition and funding authorities—can achieve near-term success if it has bipartisan support from Congress. Yet MDA has struggled to develop and deploy systems that

---

[149] James S. Lay, "Memorandum for the National Security Council: U.S. Policy on Continental Defense," Washington, D.C.: White House, 1960.

[150] Dabrowski, 2013.

[151] Ronald Reagan, "Address to the Nation on Defense and National Security," Washington, D.C., March 23, 1983.

[152] Dabrowski, 2013.

meet the requirements from the individual services because of the high degree of separation from the services. For example, the Ground-Based Midcourse Defense system has proven to be somewhat unreliable, and MDA has struggled to deploy other systems that ostensibly should have already been sent to the services.

MDA has also struggled to adapt to threat picture changes because of its focus on ballistic missiles and budget constraints. MDA is focused on the ballistic missile threat. However, with the development of hypersonic platforms and nuclear-tipped cruise missiles by China and Russia, ballistic missiles may soon be of only secondary concern. Further, MDA's budget is being progressively tied up in procurement and operations and maintenance costs for systems that should be operated by military services. This decreases the amount of funding MDA can use to research mitigating technology for the above emerging threats.

The future Space Force will need to have a mission broad enough to adjust to future space-based technologies. If the Space Force mission were too narrow, then it would be outpaced by changing technology and would be unable to add to the capabilities of the joint force.

Space plays an important supporting role in joint operations: long-range precision-guided munitions require PNT data to reach their targets, ships require communications satellites to speak to land-based forces and each other, and so forth. If the Space Force were too removed from the needs of the other services, and the Air Force in particular, then its budget would constantly be under siege from the larger services.

National Reconnaissance Office

Justification for Creation
The justification for creating NRO was founded on two Cold War–era facts. First, the U.S intelligence community had limited information about the size, shape, and capabilities of the Soviet military. Second, U.S. policymakers were deeply concerned about a surprise strategic attack by the Soviet Union.[153] With the advent of aircraft and satellite-based ISR capabilities in the late 1950s, image intelligence and measurement and signature intelligence became critical to the United States' ability to collect information about the Soviet Union.[154] When Soviet SA-2 surface-to-air missiles downed a CIA U-2 in 1960, the United States became reliant on space-based assets to gather intelligence.[155]

---

[153] Bruce Berkowitz and Michael Suk, *The National Reconnaissance Office at 50 Years: A Brief History*, Chantilly, Va.: Center for the Study of National Reconnaissance, National Reconnaissance Office, July 2018, pp. 1–2.

[154] Berkowitz and Suk, 2018, pp. 1–2.

[155] Laurie, 2001, p. 8.

During this same period, U.S. space-based reconnaissance efforts were spread across the military services and the CIA in stovepipe projects. Although the Navy's Galactic Radiation and Background (GRAB) and the CIA's CORONA programs were successful, they were also limited in scope.[156] In response to the above factors, on September 6, 1961, the John F. Kennedy administration created NRO to consolidate national space-based reconnaissance assets under a civilian organization to reduce inefficiencies from duplication of effort across the military services.[157] NRO was also created to reduce perceived leaks from congressional oversight by using CIA "unvouchered" funding instead of DoD's more tightly monitored spending.[158]

History

At the start of the Cold War, U.S. policymakers were concerned with the possibility of being strategically surprised by a Soviet attack.[159] Soviet control over information distribution, even for mundane things, such as telephone books, created a knowledge vacuum in the U.S. government about Soviet plans and capabilities.[160] In the short term, the United States turned to high-altitude flyovers of Soviet territory with the U-2. Overflight operations carried significant risks, however, as penetrating Soviet airspace might have constituted an act of war.[161] Therefore, President Dwight Eisenhower organized a blue-ribbon panel to research technology-based solutions for strategic warning in 1954.

The Technological Capabilities Panel found that a bureaucratic drag within the CIA and armed services had slowed the development of intelligence satellite programs GRAB (administered by the Navy) and Satellite and Missile Observation System (SAMOS; administered by the Air Force). The panel concluded that a new classified civilian organization was required.[162] The panel recommended that this new organization should bring together the entire satellite life cycle: design, acquisition, and operations.[163]

The executive branch did not fully implement the Technological Capabilities Panel's recommendations until after the downing of a CIA U-2 (flown by Francis Gary Powers) over Soviet territory in May 1960. In addition to the ensuing diplomatic

---

[156] Laurie, 2001, pp. 8–9, citing DoD, *Responsibility for Development of Space Systems*, Washington, D.C., Department of Defense Directive 5160.32, March 6, 1961.

[157] Berkowitz and Suk, 2018, pp. 10–12.

[158] Laurie, 2001, p. 2.

[159] Berkowitz and Suk, 2018, pp. 1–2.

[160] Berkowitz and Suk, 2018, p. 2.

[161] Laurie, 2001, p. 8.

[162] Berkowitz and Suk, 2018, p. 1.

[163] Berkowitz and Suk, 2018, pp. 1, 9.

standoff, the downing of the U-2 left the United States blind to Soviet actions behind the Iron Curtain. [164] In August 1960, Eisenhower directed the creation of the Office of Missile and Satellite Systems under the secretary of the Air Force to speed the development of SAMOS. [165]

The Kennedy administration further considered satellite-based intelligence gathering in 1961 by forming NRO. [166] The newly made organization took control of the CIA's GRAB, the Navy's CORONA, and the Air Force's SAMOS programs. Secretary of Defense Robert McNamara's intent was for NRO to independently develop, procure, and operate space-based intelligence assets using CIA unvouchered funding. However, organizational interests from the services and the CIA undermined this position. [167] Unlike a fully independent civilian agency, NRO was directed by a senior DoD official, whose deputy was a career CIA officer. [168] The NRO functional organs were based on the programs it inherited: Program A handled Air Force satellite reconnaissance. Program B did the same for the CIA, Program C did the same for the Navy, and Project D did the same for the joint Air Force–CIA aerial mission. In practice, these offices acted as a federation of independent projects, developing individual work cultures based on the organizations they supported (and generally came from), instead of as a cohesive whole. [169] Although it was not the integrated office the Kennedy administration had hoped for, NRO did prove successful at developing and employing space-based intelligence assets, but not without internal wrangling between the CIA, Air Force, and Navy over funding and control.

This organizational structure would not appreciably change until 1991. However, over the same period, NRO began to take on more responsibilities, as both a satellite operator and an intelligence collector. During the Gulf War, NRO was criticized for the slow pace of overhead intelligence. War planners argued that outdated or misleading reports led to several missed opportunities in the field and that a more responsive and real-time system was required. [170]

A commission appointed by then–Director of Central Intelligence Robert Gates found that NRO's "alphabet organization" created redundancy of effort across the organization, which could be solved through reorganization and consideration. [171]

---

[164] Laurie, 2001, p. 8.

[165] Laurie, 2001, p. 9.

[166] Laurie, 2001, p. 10.

[167] Laurie, 2001, pp. 7–10.

[168] Berkowitz and Suk, 2018, p. 13.

[169] Berkowitz and Suk, 2018, pp. 13–14.

[170] Laurie, 2001, p. 7.

[171] Berkowitz and Suk, 2018, pp. 23–24.

Then–NRO Director Martin Faga accepted the panel's and other recommendations,[172] resulting in President George H. W. Bush signing National Security Directive 67.[173] The directive reorganized NRO into three functions-based directorates: Signals Intelligence Systems Acquisition and Operations, Imagery Intelligence Systems Acquisition and Operations, and Communications.[174]

Implications for the Space Force

NRO serves as an important source of lessons for developing the future Space Force, as it both directly deals with the battle domain and shares an organizational history with the Air Force.

First, both personnel and management independence are crucial for organizational independence. As noted, NRO operated as a loose collection of separate projects instead of as a unified office until 1991. The organizational design was based off the programs it had inherited, its personnel were primarily members of the CIA or the Air Force, and its leadership came from DoD and the intelligence community. Even in the years after the 1991 redesign, NRO has continued to be primarily staffed by personnel from other organizations, leaving it open to disruption from interoffice budget fights.

The second lesson is that a new organization needs clear policy jurisdiction. Although NRO's part-defense, part-intelligence roll might have allowed it to remain a closely guarded secret until the late 1980s, it has also led to conflict between DoD and the CIA over policy jurisdiction. Wrangling between the two over leadership roles and relationships inside NRO has at times disrupted operations. The future Space Force will require a clearly worded policy jurisdiction in its Title 10 authorities.

Global Strike Command

Background

Following the end of the Cold War and throughout the 1990s, the emphasis on the United States' nuclear arsenal and the role of nuclear deterrence in U.S. national security declined dramatically. The global geopolitical and threat environment had changed, and, as a result, national defense priorities were changing. Significant defense downsizing was under way. The size of the nuclear arsenal was reduced, and funding for the nuclear enterprise (including in the Department of Energy and in DoD) decreased from the post–Cold War levels. The Air Force was making numerous insti-

---

[172] Similar recommendations were from another panel that Faga himself had commissioned.

[173] White House, *Intelligence Capabilities: 1992–2005*, Washington, D.C., National Security Directive 67, March 30, 1992.

[174] Berkowitz and Suk, 2018, pp. 23–24.

tutional and organizational changes and shifting its emphasis to conventional missions and expeditionary operations.[175]

One consequence of this strategic shift was the decommissioning of Strategic Air Command in 1992. Components of the strategic nuclear strike forces were dispersed among ACC (long-range bombers and ISR assets), AFSPC (ICBMs), and Air Mobility Command (tankers). The need for nuclear deterrence, however, remained, as did the need for the "zero defects" culture that is required for safe, secure, and effective nuclear operations. But as a result of the disaggregation and the reductions in manpower, funding, and focus, the nuclear enterprise's capabilities began to atrophy.

By 2007, it was becoming clear that these factors were resulting in major problems in the conduct of nuclear operations and related activities.

Incidents, Investigations, Reports, and Reinvigoration

Grave and far-reaching concerns arose over incidents in 2006 and 2007. In two separate incidents in October and in November of 2006, critical nuclear-related ICBM parts were inadvertently shipped to Taiwan. In August 2007, a B-52 flew from Minot Air Force Base, North Dakota, to Barksdale Air Force Base, Louisiana, with the crew unaware that six live nuclear warheads were aboard.[176]

These events were covered widely by the media. Importantly, they served as a watershed for the decisions and changes that would soon follow. First, the discovery in June 2008 of the fuses in Taiwan (shipped in 2006) led Secretary of Defense Gates to relieve both the chief of staff of the Air Force and the secretary of the Air Force on the same day.[177]

Second, these two events triggered a series of studies and reports that provided insights into declining mission capability and ways to address the multitude of problems related to the nuclear enterprise. A task force led by James R. Schlesinger was appointed, and several other investigations and investigative commissions followed. Although the hundreds of findings and recommendations that resulted are beyond the scope of this report, we include two findings. Two reports from Schlesinger's task force, among others, underscored the finding that cuts had gone too far, and the Defense Science Board concluded: "Since the end of the Cold War, there has been a marked decline in the level and intensity of focus on the nuclear enterprise and the nuclear mission. The decline in focus took place gradually as changes were made to policies, proce-

---

[175] Ross, 2012.

[176] Ross, 2012; Jim Garamone, "Air Force Global Strike Command Will Stress Nuclear Mission," *American Forces Press Service*, August 7, 2009.

[177] Robert M. Gates, *Duty: Memoirs of a Secretary at War*, New York: Vintage, 2015.

dures, and processes."[178] The reviews and their subsequent reports clearly put forward and outlined the need for a reinvigoration of the nuclear enterprise.[179]

Lastly, the release of this series of reports, starting in 2008, spurred swift action to reinvigorate the nuclear enterprise. The secretary of the Air Force and chief of staff of the Air Force established the nuclear enterprise's reinvigoration as a top priority. They directed the stand-up of AFGSC, a dramatic expansion of the Air Force Nuclear Weapons Center, and a nuclear-focused Air Staff directorate (AF/A10 [Strategic Deterrence and Nuclear Integration]). Many of these structural changes were implemented in less than seven months.

Stand-Up of Global Strike Command

To address concerns about the stewardship of the U.S. nuclear arsenal, the Air Force created a new command.[180] AFGSC—created to provide combat-ready forces to conduct strategic nuclear deterrence and global strike operations—was activated on August 7, 2009. The command was installed at Barksdale Air Force Base, home of the 8th Air Force, with Lt Gen Frank G. Klotz as its first commander. Responsibility for nuclear-capable assets of AFSPC transferred to Global Strike Command in December 2009 and the nuclear-capable assets of ACC in February 2010. Global Strike Command was initially tasked with responsibility for all long-range nuclear-capable bombers (the B-52H and B-2) and ICBMs (Minuteman III); it later gained responsibility for all long-range bombers, including the B-1.[181]

The reinvigoration of the nuclear enterprise required additional manpower. Approximately 2,500 billets were added to the nuclear enterprise, over and above the 900 new billets required to stand up AFGSC. About 900 of the 2,500 were allocated to the 4th B-52 squadron and B-52 maintenance capability at Minot Air Force Base. Three hundred sixty positions were added to support ICBM operations. The remaining 1,240 were used to fill needs across the enterprise. Also, key nuclear billets were established to further identify positions that required greater depth of nuclear expertise and experience, and special experience identifiers were developed to identify positions that required special experience and to identify personnel who possessed that experience. It was understood that Global Strike Command's headquarters would take several years to fully man. Temporary duty personnel supplemented the initial permanent

---

[178] Defense Science Board Permanent Task Force on Nuclear Weapons Surety, *Report on the Unauthorized Movement of Nuclear Weapons*, Washington, D.C.: Office of the Under Secretary of Defense for Acquisition, Technology, and Logistics, February 2008.

[179] Ross, 2012.

[180] For example, see Thomas E. Ricks and Joby Warrick, "Tough Punishment Expected for Warhead Errors; Officers May Lose Commands After Nuclear Missiles Were Flown on Bomber," *Washington Post*, October 18, 2007.

[181] AFGSC, Public Affairs Office, "AFGSC Mission and Vision," webpage, undated; AFGSC, "Air Force Global Strike Command," fact sheet, 2015.

party that stood up the headquarters, and contractors were also added, many on two-year contracts.

Decisions at Stand-Up and Insights of Relevance

The stand-up of Global Strike Command is likely one of the best cases to study, vis-à-vis the stand-up of the Space Force, because of the similarities in the parent structure and the rationale for the new organization, and also because the Global Strike Command case is so recent. The Air Force that gave rise to AFGSC is essentially the same organization and structure as today's Air Force, and Global Strike Command had to be carved out of what is essentially today's existing resources and manpower. (In contrast, the Air Force's separation and transition from the Army at the end of World War II had the advantage of having tens of thousands of combat-experienced Army Air Forces personnel from which to draw.) Moreover, the rationale for the stand-up of the Space Force is similar to the motivations to create Global Strike Command: realization that the current structure of Air Force is not able to facilitate the need for specific, enhanced strategic and operational capabilities. Additionally, lessons learned from Global Strike Command's stand-up are still fresh, and those who were there at the stand-up are available to speak authoritatively not only to the stand-up itself but also to the critical insights they have gained in the intervening years since 2008. In this section, we will examine some of those insights.

Make the Tough Decisions Up Front to Support Mission Clarity and Integrity

Decisions to pull components and resources from one organization and award them to another is always difficult, and such moves are fraught with political skirmishes, power struggles, and lack of cooperation—or even, at times, goodwill. However, mission clarity and mission integrity must be foremost considerations, driving the process of deciding what goes where and when. According to our subject-matter experts, establishing Global Strike Command's roles and missions resulted in a "running gun battle." Functions moved to Barksdale Air Force Base, but the people manning them did not. Disputes with ACC were not resolved for years. All missiles and bombers (nuclear and conventional) moved to AFGSC, but very few billets accompanied the transfer of B1s and E4. Major operational testing and evaluation capabilities did not transfer to Global Strike Command. Instead, these were kept in ACC and AFSPC, which contributed to the underresourcing.

As one senior leader who was central to the Global Strike Command stand-up advised, "Don't count on evolution to solve the political problems. Make the tough decisions up-front—go big early."

Strive for Clarity in All Authorities

Lt Gen Frank Klotz, the first AFGSC commander, directed early in the stand-up that the new command would not rely on existing ACC and AFSPC instructions to guide its activities but would write its own documents—about 200 in all. He had found a basic lack of understanding among nuclear airmen of the roles, responsibilities, and authorities for conducting nuclear deterrence operations, following two decades of reduced resources, lack of leadership attention, and the dispersion of responsibilities among multiple major commands. A well-grounded understanding of who had what authorities and what those authorities entailed was, he believed, fundamental to the establishment of the new command in all areas—from operations and maintenance to personnel and public affairs. Moreover, drafting its own guidance gave AFGSC the opportunity to change what needed to be changed and to consciously and deliberately eliminate those activities that were no longer of value. Although the task was daunting and time-consuming, Klotz related that it also proved to be intrinsically motivating for AFGSC airmen at the time—it conveyed the certainty that they were having an important and discernible impact and were setting a precedent that would last for years.

Ensure Parity in Rank from the Beginning

When AFGSC was initially stood up, an emphasis on creating parity with other major commands was not deemed essential. As a result, the AFGSC commander was designated a three-star, in part based on the Air Force Special Operations Command model. But the Air Force Special Operations Command model was perhaps not the optimal choice; its commander was or is the same rank as other SOCOM component commanders; thus, the commander would have parity in terms of influence, decision-making, and competition for resources with other SOCOM components. The AFGSC commander, however, was a three-star, competing for influence and resources with the other major command commanders—a different arena in which all colleagues had four stars. Additionally, four-star general officers have meetings in which three-stars are not present and forums in which three-stars cannot participate. The possibility always existed that discussions leading to important decisions (e.g., senior officer personnel decisions) would take place in these sessions—effectively precluding any direct input from the AFGSC commander or any representation from the new major command.

AFGSC operated without a four-star commander for several years, and many feel that this initial decision handicapped AFGSC for several years. The A10, as well, was arguably handicapped at the Air Staff as a two-star, when all other two letters were and are three-stars. This situation persisted for several years.

Fortify the Money

Ensuring sufficient resources to stand up a new organization is always a challenge. The Air Force's top-line budget generally grows only incrementally, and manpower levels are even harder to increase. Thus, carving out the money and people necessary to establish a new major command is essentially a zero-sum game—funds for the new outfit generally must come from existing units, or that at least is what the leaders of those organizations often perceive.

The initial leadership cadre of AFGSC generally felt that it had adequate financial resources from the start. Support from the secretary and chief of staff of the Air Force was strong, and the quarterly Nuclear Oversight Board meetings kept DAF senior leaders focused on the needs of reinvigorating the Air Force nuclear enterprise. That said, the Air Force corporate process drives budget decisions, and as funding issues move from panel to board to council, many trade-offs are made. AFGSC representatives frequently encountered resistance at their level to the amount of funding that was going to particular nuclear enterprise needs. Moreover, AFGSC might have been somewhat at a disadvantage in this annual tug and pull, since the command was still in the process of building up its program and budget expertise and having to deal with more-experienced practitioners from the other major commands and mission areas. The AFGSC leadership team paid careful attention to the budget-making process during the command's initial years and did not hesitate to engage directly with the secretary and chief when it deemed necessary. That high-level top cover was absolutely essential to the success of establishing a new command.

Fight for the Proper Manpower Up Front or Live with the Consequences Indefinitely and Perhaps Permanently

Manpower was a different issue. When the command initially stood up in August 2009, only 40–50 of the expected 800-plus permanent party personnel were actually on board at the Barksdale Air Force Base headquarters. As noted, key functions had to be performed by temporary duty detailees, contractors, and other personnel "on loan" to AFGSC. Recruiting the initial cadre of headquarters personnel was facilitated by the desire of many nuclear enterprise "veterans" to help establish a command focused on nuclear operations or by others who relished the opportunity to be in on the ground floor of a new, high-visibility organization. Rightly or wrongly, several early AFGSC leaders believe that the command was short-changed in terms of manpower from the outset. Not all the manpower slots in ACC and AFSPC that dealt with bomber or missile-related issues (especially in functions like requirements and financial management) were transferred over to AFGSC. Attempts to appeal directly to the Air Force manpower directorate at Randolph Air Force Base for additional manpower slots bore little fruit.

The consequences have bedeviled the command ever since. It remains questionable whether AFGSC today has sufficient people to deal with the extraordinary challenges posed by having to field, operate, and maintain multiple new weapon systems and capabilities as part of the United States' nuclear modernization program.

Beyond Manning, Flexibilities Needed in Human Capital Policies and Processes

Whenever broad structural and organizational changes take place, human capital consequences are sure to follow. Serious, systemic human capital issues had languished, unaddressed, in the nuclear career fields in the years of dispersed nuclear missions, leading up to the incidents in 2006 and 2007, and they continued throughout the first few years after the stand-up of the AFGSC. The problems were acute—the 898th Munitions Squadron was decertified in 2010, maintenance units in particular were plagued with operational distress, officer missile career fields had unsustainable billet structures, the failures in all types of inspections were unremitting, and the Air Force's enlisted assignment system could not place the most-qualified nuclear senior noncommissioned officers where they were critically needed.[182]

In the fall of 2009, Maj Gen Don Alston (A10) commissioned the first human capital strategy for the nuclear enterprise, an effort that continued until the end of 2012. The strategy first undertook a diagnosis of the most-critical manpower, personnel, and training issues. Six broad, strategic human capital issues were elevated that required the approval of the chief of staff and secretary of the Air Force to address. Another critical finding to emerge was that a subset of core enlisted nuclear career fields—2W2, 1C3, and 3P0—was in trouble and unlikely to recover without focused intervention. Solutions produced by the Human Capital Strategy involved critically needed changes to the nuclear senior enlisted assignment system to ensure the best quality, expediting a set of urgently needed manpower studies across the nuclear enterprise, a rework of the billet structure in several key nuclear enterprise career fields, and an eventual complete revamp of the Personnel Reliability Program administrative process (which had been significantly affecting available manning).[183]

Large-scale structural change cannot be accomplished in a vacuum, and the stand-up of the Space Force will create challenges and burdens on the existing personnel, manpower, and training and education policies and processes. The nuclear enterprise had to undertake an organized effort and then escalate the need for changes and flexibilities to DAF senior leadership to prevail. We expect the Space Force to also have to lobby hard with DAF and DoD senior leadership for the flexibilities in the human capital policies and processes that it would undoubtedly require. A planned human capital strategy and implementation could diagnose and implement the changes in personnel policies and processes that would be needed—and would head off the problems—to keep pace with the new, broad structural changes and could be a radical departure from the current organizational and mission status quo.

---

[182] Ross, 2012.

[183] Ross, 2012.

Career Field Sustainability: Acquisition Officers

We use our methodology to examine two additional Air Force career fields of high importance to the Space Force: 62E (Developmental Engineer) and 63A (Acquisition Manager), again considering the three criteria of *pyramid health, career path viability,* and *leadership opportunities.* These two career fields would likely be the second and third largest officer career fields in the Space Force, just behind the 13SX career field in size.[1] See Chapter Five for a full explanation of the career field sustainability model and methodology.

These two career fields are uniquely intertwined with regard to career progression and leadership opportunities. They also have additional complexities. The analyses for them must be run independently, but also comingled, to assess their potential to be Space Force autonomous, the Air Force with integrated space track, or the Air Force with Space Force assignments.

These career field analyses are preliminary, and they will need to be followed by more in-depth analyses once the structure, assets, and resources of the Space Force are more definitively known.

Acquisition Workforce Overview

The acquisition function would be a core function of the Space Force, and the acquisition workforce would be critical for the successful procurement of space systems. The Air Force acquisition workforce is composed of five distinct career fields, each with unique skill sets, specialized development paths, and senior leadership opportunities. The 2019 *Air Force Officer Classification Directory* identifies the five Air Force specialties in the Acquisition and Financial Management career area:[2]

- Scientist (61XX)

---

[1] Royx, 2019.

[2] Air Force Personnel Center, *Air Force Officer Classification Directory*, Randolph Air Force Base, Tex., April 30, 2019.

- Developmental Engineer (62XX)
- Acquisition Manager (63XX)[3]
- Contracting (64XX)
- Finance (65XX).

Most program offices, including space system program offices, require a mix of all five specialties to manage procurement. However, at this time, our study considers only the 62XX and 63XX career fields for potential transfer to the Space Force. As mentioned, these two career fields are uniquely intertwined in both career progression and leadership opportunities.

Developmental Engineer (62E) Career Field Assessment

The engineering discipline in the Air Force is categorized as a portion of the acquisition workforce. In developing a specialty code structure for managing this career field, the Air Force employs a set of suffixes. As seen in Table B.1, six of the eight suffixes have academic degree requirements for award of the AFSC. The *flight test* suffix (*F*) must be earned through completion of the Air Force Flight Test Engineer Course, and the *project* suffix (*G*) represents a catchall category for officers with broad engineering degrees and for billets with nonspecific or cross-disciplinary needs.

Currently, AFSPC has more than 600 developmental engineering billet authorizations. The largest single group of these are ascribed to project engineers (*G* suffix, 247 billets); as a result, when combined with the set of billets where no suffix is indicated, nearly two-thirds of AFSPC engineering billets require no specific academic background. By comparison, just over 11 percent of the billets explicitly require space system engineering expertise. Without further analysis, we cannot know whether this is a true characterization of positional duties, whether it is a reflection on the scarcity of qualified candidates for such positions, or whether it is a flexibility that has been built into the AFSPC workforce to move engineers across billets, as needed. Further analysis will be critical to assessing whether portions of the Developmental Engineer career field should transition wholly to the new Space Force as an autonomous specialty or whether the Air Force with integrated space track would be an acceptable alternative.

Pyramid Health

A healthy career field pyramid has a sufficient pool of candidates in any given rank to support competitive selection for the next rank, while at the same time avoiding

---

[3] The 63AX AFSC is formally called Acquisition Manager, which could be confused with references to the broader set of specialties collectively referenced as the *acquisition workforce*. As a result, 63XX officers are frequently called *program managers* for clarity. Accordingly, this report will refer to 63XX officers as *program managers*.

Table B.1
Air Force Space Command 62E Authorizations (as of March 31, 2019)

Suffix	Discipline	LT	Capt	Maj	Lt Col	Col	Total
None	—	17	59	74	30	2	182
A[a]	Aeronautical	9	12	4	—	—	25
B[a]	Astronautical	37	32	4	2	—	75
C[a]	Computer Systems	6	5	—	—	—	11
E[a]	Electrical/Electronic	42	51	9	1	—	103
F	Flight Test	1	—	3	1	—	5
G	Project	63	119	52	9	4	247
H[a]	Mechanical	7	15	—	—	—	22
I[a]	Systems/Industrial/ Human Factors	—	—	—	—	—	0
Total		182	293	146	43	6	670

[a] The initial AFSC award requires an undergraduate engineering degree in that discipline.

overmanning, which can lead to depressed promotion potential. Table B.2 shows the overall LAF promotion selectivity. Comparatively, Table B.3 shows the corresponding promotion rates for all 62XX officers, and Table B.4 shows the rates for astronautical developmental engineers. Promotion ratios are an important consideration in pyramid health.

The column on selectivity to higher grade at the far right of each table represents the number of candidates available at the lower grade for each open billet at the next rank (assuming that each rank is 100 percent manned). As seen in Table B.2, there are three O5s available for each O6 position in the overall LAF.

Table B.2
Line of the Air Force Promotion Opportunities, by Rank

Grade	Authorizations (as of March 31, 2019)	Assumed Time in Grade (years)	Promotion	Promotions per Year	Number Separated per Year	Selectivity from Lower Grade	Selectivity to Higher Grade
O1, O2, O3	25,227	10	To O1, O2, O3	2,522.7	—	—	1.2
O4	10,156	5	O3 to O4	2,031.2	491.5	1.2	1.6
O5	7,449	6	O4 to O5	1,241.5	789.7	1.6	3.0
O6	2,447	6	O5 to O6	407.8	833.7	3.0	—

Table B.3
Air Force Space Command 62XX Promotion Opportunities, by Rank

Grade	Authorizations (as of March 31, 2019)	Assumed Time in Grade (years)	Promotion	Promotions per Year	Number Separated per Year	Selectivity from Lower Grade	Selectivity to Higher Grade
O1, O2, O3	475	10	To O1, O2, O3	47.5	—	—	1.6
O4	146	5	O3 to O4	29.2	18.3	1.6	4.1
O5	43	6	O4 to O5	7.2	22.0	4.1	7.2
O6	6	6	O5 to O6	1.0	6.2	7.2	—

By comparison, as seen in Table B.3, there are more than seven O5 developmental engineers in AFSPC for every O6, which is not an acceptable ratio based on our assumption that any set of promotion ratios must be equitable with the Air Force's (see Chapter Five). Table B.4 reveals no promotion opportunity for AFSPC astronautical engineers, because of the absence of O6 billets specifically requiring the *astronautical engineering* suffix (*B*). Similarly, the promotion ratios for O4s to O5 in both tables are lower than the current LAF ratios.

Taken by themselves, these data would suggest that, if the Air Force 62E career field moved over "as is," the Space Force 62E career field would have downward pressure on O5 and O6 promotion rates, compared with LAF promotion rates. Some of this downward pressure could be mitigated by the undermanned status of the career field. However, recalling that 62XX officers migrate to 63XX positions at higher ranks, in the next section we will examine the combined 62XX/63XX career fields.

Table B.4
Air Force Space Command 62XXB Promotion Opportunities, by Rank

Grade	Authorizations (as of March 31, 2019)	Assumed Time in Grade (years)	Promotion	Promotions per Year	Number Separated per Year	Selectivity from Lower Grade	Selectivity to Higher Grade
O1, O2, O3	69	10	To O1, O2, O3	6.9	—	—	8.6
O4	4	5	O3 to O4	0.8	6.1	8.6	2.4
O5	2	6	O4 to O5	0.3	0.5	2.4	0.0
O6	0	6	O5 to O6	0.0	0.3	—	—

Career Path Viability

The career path progression for developmental engineers focuses on diversity of experiences in the early assignments. Early company grade officer assignments may be in acquisition, operations, or research and development. Alternately, junior officers may be selected for advanced academic degree programs through the Air Force Institute of Technology.

For field grade officers, the focus turns to selection for leadership positions consistent with an officer's time in grade. Board selection for materiel leader positions (typically an O5 opportunity) is one such example. Additionally, field grade officers can be expected to serve on major command staffs, the Air Staff, or the Joint Staff.

In assessing the potential to shift the Air Force 62XX career field to the Space Force, the primary consideration is maintaining this diversity of opportunities. Figure B.1 shows the current distribution of developmental engineers within the Air Force.[4] Locations listed in red at the left of the figure represent those bases that would have (in whole or in part) units that would become Space Force units. Locations in black have units with at least a portion of a space mission (such as the National Air and Space Intelligence Center at Wright-Patterson Air Force Base) but that are unlikely to become Space Force organizations. The diversity of red- and black-labeled locations gives the initial indication that there would be sufficient career path opportunities for Space Force developmental engineers. But, as stated above, when considering pyramid health, without deeper insight into the specialties required of engineers at these locations, it is difficult to make more than a preliminary assessment.

Leadership Opportunities

A major confounding factor in assessing senior leadership opportunities for 62XX officers in the proposed Space Force is the interplay between the 62XX and 63XX career fields at the senior ranks. Through design or evolution, these two specialties have interwoven paths of upward mobility, and neither is self-sustaining in senior leadership positions. For engineers, the number of senior positions available (62S0 Materiel Leader billets and O6 62EX billets) is sparse compared with the number of lower-echelon billets, implying low promotion opportunities late in one's career. For program managers, available billets at the junior officer level do not create a sufficiently deep and broad pool of potential candidates for senior positions. However, by fusing the two fields together, they are mutually supportive for growth. This factor would be important to career field migration to the Space Force, and it is considered more deeply below.

---

[4] This figure was created based on O6 positional data; however, it is reasonable to extrapolate the presence of fewer senior officers in these locations as well.

Figure B.1
Current Air Force 62XX Assignment Distribution

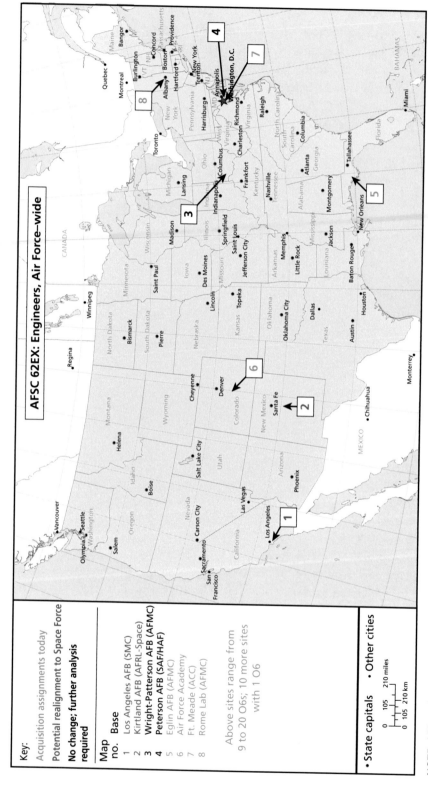

AFSC 62EX: Engineers, Air Force–wide

Key:

Acquisition assignments today

Potential realignment to Space Force

No change; further analysis required

Map no.	Base
1	Los Angeles AFB (SMC)
2	Kirtland AFB (AFRL-Space)
3	Wright-Patterson AFB (AFMC)
4	Peterson AFB (SAF/HAF)
5	Eglin AFB (AFMC)
6	Air Force Academy
7	Ft. Meade (ACC)
8	Rome Lab (AFMC)

Above sites range from 9 to 20 O6s; 10 more sites with 1 O6

• State capitals   • Other cities

0   105   210 miles

0   105   210 km

NOTE: AFB = Air Force base; AFMC = Air Force Materiel Command; SMC = Space and Missile Systems Center; SAF = secretary of the Air Force; HAF = Headquarters, Air Force.

Program Manager (63A) Career Field Assessment

Pyramid Health

Table B.5 shows the promotion ratios for 63XX officers. We find there are 0.8 captains for each O4 billet, reinforcing the need for 62XX officers to cross-flow into the 63XX career field. Pyramid strength for 63XXs from O4 to O5 (1.7) and O5 to O6 (3.5) very closely matches the LAF assessment (1.6 and 3.0, respectively), which is a positive outcome.

Table B.6 further considers the combined 62XX and 63XX career fields. The combination of AFSCs reveals balance in both junior and senior ranks. Selectivity to O4 is greater than 1.0, signaling a sustainable base to the pyramid. And selectivity to O5 and O6 (2.2 and 3.9) is reasonably close to the overall LAF numbers from Table B.2.

Career Path Viability

Similar to the 62XX career field, the 63XX career field seeks a broad base of experiences for company grade officers in their first decade of assignments (see Figure B.2). Exam-

Table B.5
Air Force Space Command 63XX Promotion Opportunities, by Rank

Grade	Authorizations (as of March 31, 2019)	Assumed Time in Grade (years)	Promotion	Promotions per Year	Number Separated per Year	Selectivity from Lower Grade	Selectivity to Higher Grade
O1, O2, O3	355	10	To O1, O2, O3	35.5	—	—	0.8
O4	214	5	O3 to O4	42.8	−7.3	0.8	1.7
O5	154	6	O4 to O5	25.7	17.1	1.7	3.5
O6	44	6	O5 to O6	7.3	18.3	3.5	—

Table B.6
Air Force Space Command Combined 62XX and 63XX Promotion Opportunities, by Rank

Grade	Authorizations (as of March 31, 2019)	Assumed Time in Grade (years)	Promotion	Promotions per Year	Number Separated per Year	Selectivity from Lower Grade	Selectivity to Higher Grade
O1, O2, O3	830	10	To O1, O2, O3	83.0	—	—	1.2
O4	360	5	O3 to O4	72.0	11.0	1.2	2.2
O5	197	6	O4 to O5	32.8	39.2	2.2	3.9
O6	50	6	O5 to O6	8.3	24.5	3.9	—

Figure B.2
Current Air Force 63XX Assignment Distribution

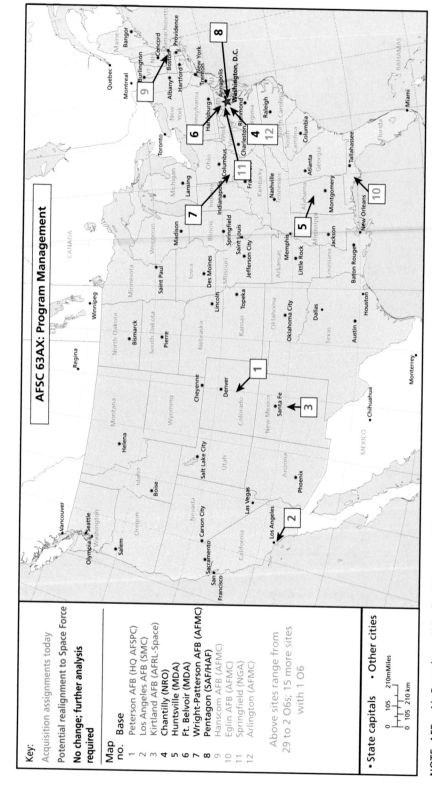

AFSC 63AX: Program Management

Key:
Acquisition assignments today
Potential realignment to Space Force
No change; further analysis required

Map no.	Base
1	Peterson AFB (HQ AFSPC)
2	Los Angeles AFB (SMC)
3	Kirtland AFB (AFRL-Space)
4	Chantilly (NRO)
5	Huntsville (MDA)
6	Ft. Belvoir (MDA)
7	Wright-Patterson AFB (AFMC)
8	Pentagon (SAF/HAF)
9	Hanscom AFB (AFMC)
10	Eglin AFB (AFMC)
11	Springfield (NGA)
12	Arlington (AFMC)

Above sites range from 29 to 2 O6s; 15 more sites with 1 O6

• State capitals   • Other cities

NOTE: AFB = Air Force base; HQ = headquarters; AFMC = Air Force Materiel Command; SMC = Space and Missile Systems Center; SAF = secretary of the Air Force; HAF = Headquarters, Air Force; NGA = National Geospatial-Intelligence Agency.

ples of broadening include operational experience tours, advanced academic degree attainment, or education-with-industry assignments. "Mainstream" program management is also a possibility, in which one gains foundational knowledge in a program office or product center. As a senior company grade officer or junior field grade officer, other program management opportunities emerge, including serving as a Defense Contract Management Agency representative in a manufacturing facility, instructing at a Defense Acquisition University campus or the Air Force Institute of Technology, or overseeing research and development at an Air Force laboratory. As officers progress to the senior field grade officer ranks, selection as a materiel leader (discussed in the next section) opens further prospects for advancement. The 63XX career field has adequate breadth and depth of assignments to create viable career paths.

Leadership Opportunities

As discussed, the 63XX career field structure at the senior levels is unsustainable without cross-flow from the 62XX career field.

Although many O6 positions for developmental engineers and program managers provide significant opportunity for the leadership, a specific set of command billets is centrally managed by a command selection board. These billets are materiel leader and senior materiel leader positions. In general terms, these are analogous to squadron, group, and wing command positions in operational Air Force units. The four distinct specialty codes for these positions are

- 60C0: Senior Materiel Leader-Upper Echelon
- 63G0: Senior Materiel Leader-Lower Echelon
- 63S0: Materiel Leader
- 62S0: Materiel Leader (Developmental Engineer).

Materiel leader billets are typically O5 branch chief positions or higher, with stated requirements for the number of direct reports, the acquisition category program designation, and funding levels of responsibility. Senior materiel leaders are O6 positions, also with specific standardized criteria.

This study considered the location and distribution of the O6 senior materiel leader billets to assess potential senior leadership opportunities in the proposed Space Force. Table B.7 reveals that AFSPC maintains nearly one-third of the designated senior materiel leader billets; Air Force Materiel Command holds nearly all other senior materiel leader billets, arrayed as shown in Figure B.3. As shown in both the table and figure, the AFSPC billets are assigned either at AFSPC bases or within units where a proposed Space Force would naturally maintain liaison units.

Figure B.3
Current Air Force Senior Materiel Leader Assignment Distribution

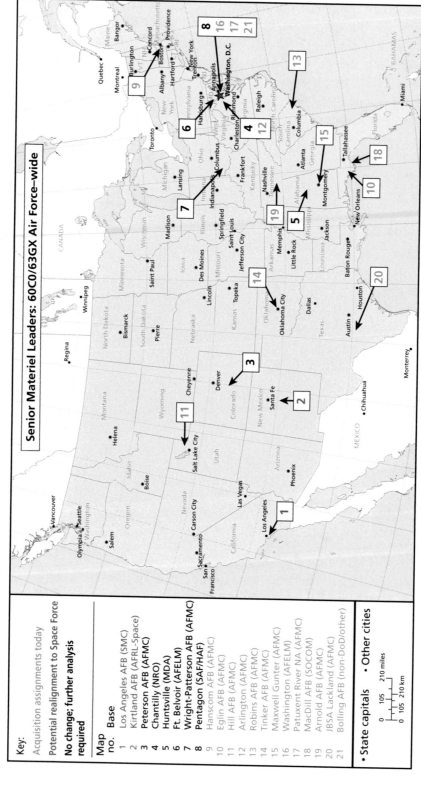

Key:
Acquisition assignments today
Potential realignment to Space Force
No change; further analysis
required

Map no.	Base
1	Los Angeles AFB (SMC)
2	Kirtland AFB (AFRL-Space)
3	Peterson AFB (AFMC)
4	Chantilly (NRO)
5	Huntsville (MDA)
6	Ft. Belvoir (AFELM)
7	Wright-Patterson AFB (AFMC)
8	Pentagon (SAF/HAF)
9	Hanscom AFB (AFMC)
10	Eglin AFB (AFMC)
11	Hill AFB (AFMC)
12	Arlington (AFMC)
13	Robins AFB (AFMC)
14	Tinker AFB (AFMC)
15	Maxwell Gunter (AFELM)
16	Washington (AFELM)
17	Patuxent River NA (AFMC)
18	MacDill AFB (SOCOM)
19	Arnold AFB (AFMC)
20	JBSA Lackland (AFMC)
21	Bolling AFB (non-DoD/other)

• State capitals   • Other cities

0   105   210 miles
0   105   210 km

Senior Materiel Leaders: 60C0/63GX Air Force–wide

NOTE: AFB = Air Force base; SMC = Space and Missile Systems Center; AFMC = Air Force Materiel Command; AFELM = Air Force Element.

Table B.7
Senior Materiel Leader Billets

AFSC	Air Force–Wide Positions	AFSPC Positions	AFSPC Locations
60C0	18	5	4: Los Angeles Air Force Base 1: Kirtland Air Force Base
63G0	94	30	17: Los Angeles Air Force Base 10: Chantilly, Va. (NRO) 3: Kirtland Air Force Base

Acquisition Workforce Conclusions

As with Air Force, the 62XX and 63XX career fields would appear to need combined management in the Space Force for both to be viable. To a greater extent than with the 13S and 14N career fields, the case of 62XX and 63XX needs further information and analysis. Information critical to a determination of whether these two could be Space Force autonomous includes, first, more-detailed information on manpower requirements for engineering billets *other than astronautical engineers* within the Space Force. We would also need to have more information about engineer and program manager acquisition billets outside AFSPC. Finally, we would need greater certainty about which billets in agencies outside the Air Force would convert to Space Force billets.

The further analysis of 62XX and 63XX is likely one of the principal tasks in the continuing work to establish the Space Force. Because this is potentially the largest contingent of officers aside from in the 13S career field, it would be a great benefit for the Space Force to establish exactly how the 62XX and 63XX career fields can be structured so as to be Space Force autonomous.

Bibliography

557th Weather Wing, "2nd Weather Squadron," webpage, undated. As of September 28, 2019:
https://www.557weatherwing.af.mil/Units/2d-Weather-Group/2d-Weather-Squadron/

Adams, Michael, "Why the OPM Hack Is Far Worse Than You Imagine," *Lawfare*, March 11, 2016. As of September 28, 2019:
https://www.lawfareblog.com/why-opm-hack-far-worse-you-imagine

AFGSC—*See* Air Force Global Strike Command.

AFRL—*See* Air Force Research Laboratory.

Air Command and Staff College, *AU-18 Space Primer*, Maxwell Air Force Base, Ala.: Air University Press, September 2009. As of September 28, 2019:
https://www.airuniversity.af.edu/Portals/10/AUPress/Books/AU-18.PDF

Air Force Global Strike Command, Air Forces Strategic–Air, "Units," webpage, undated. As of August 28, 2019:
https://www.afgsc.af.mil/units/

Air Force Global Strike Command, Public Affairs Office, "AFGSC Mission and Vision," webpage, undated. As of August 28, 2019:
https://web.archive.org/web/20090525053347/http://www.afgsc.af.mil/main/welcome.asp

————, "Air Force Global Strike Command," fact sheet, 2015. As of August 28, 2019:
https://www.af.mil/About-Us/Fact-Sheets/Display/Article/104462/air-force-global-strike-command/

Air Force Magazine, "The Air Force in Facts & Figures," in special issue, "2018 USAF Almanac," Vol. 100, No. 6, June 2018, pp. 36–57.

Air Force Personnel Center, *Air Force Officer Classification Directory*, Randolph Air Force Base, Tex., April 30, 2019

Air Force Research Laboratory, "Air Force Research Laboratory Directed Energy Directorate Air Force Maui Optical and Supercomputing (AMOS) Site," fact sheet, 2017. As of September 28, 2019:
https://www.kirtland.af.mil/Portals/52/documents/AMOSFactsheet.pdf

Alkire, Brien, Sherrill Lingel, Caroline Baxter, Christopher M. Carson, Christine Chen, David Gordon, Lawrence M. Hanser, Lance Menthe, and Daniel M. Romano, *Command and Control of Joint Air Operations in the Pacific: Methods for Comparing and Contrasting Alternative Concepts*, Santa Monica, Calif.: RAND Corporation, RR-1865-AF, 2018. As of November 13, 2019:
https://www.rand.org/pubs/research_reports/RR1865.html

Aquilani, B., C. Silvestri, A. Ruggieri, and C. Gatti, "A Systematic Literature Review on Total Quality Management Critical Success Factors and the Identification of New Avenues of Research," *TQM Journal*, Vol. 29, No. 1, 2017, pp. 184–213.

Balanced Scorecard Institute, "Balanced Scorecard Basics," webpage, undated. As of August 30, 2019:
https://www.balancedscorecard.org/BSC-Basics/About-the-Balanced-Scorecard

Barnes, Julian E., and Peter Spiegel, "Air Force's Top Leaders Are Ousted," *Los Angeles Times*, June 6, 2008.

Bartlett, Merrill L., and Jack Sweetman, *Leatherneck: An Illustrated History of the U.S. Marine Corps*, Annapolis, Md.: Naval Institute Press, 2008.

Berkowitz, Bruce, and Michael Suk, *The National Reconnaissance Office at 50 Years: A Brief History*, Chantilly, Va.: Center for the Study of National Reconnaissance, National Reconnaissance Office, July 2018.

Bowman, Martin W., *The USAAF Handbook: 1939–1945*, New York: Stackpole Books, 1997.

Builder, Carl H., *The Masks of War: American Military Styles in Strategy and Analysis*, Baltimore, Md.: Johns Hopkins University Press, 1989.

Bull, Hedley, *The Anarchical Society: A Study of Order in World Politics*, New York: Columbia University Press, 1995.

Camm, Frank, Lauren Caston, Alexander C. Hou, Forrest E. Morgan, and Alan J. Vick, *Managing Risk in USAF Force Planning*, Santa Monica, Calif.: RAND Corporation, MG-827-AF, 2009. As of November 14, 2019:
https://www.rand.org/pubs/monographs/MG827.html

Carter, Ash, *A Lasting Defeat: The Campaign Against ISIS*, Cambridge, Mass.: Belfer Center for Science and International Affairs, Harvard Kennedy School, October 2017. As of November 15, 2019:
https://www.belfercenter.org/LastingDefeat

Carton, Robert B., and Charles W. Hofer, *Measuring Organizational Performance: Metrics for Entrepreneurship and Strategic Management Research*, Cheltenham, UK: Edward Elgar Publishing, 2006.

Caton, Jeffrey L., "Joint Warfare and Military Dependence on Space," *Joint Forces Quarterly*, Winter 1995–1996, pp. 48–53.

Cebrowski, Arthur K., and John J. Garstka, "Network-Centric Warfare: Its Origin and Future," *Proceedings*, Vol. 124, January 1998.

Chaplain, Cristina T., "Space Acquisitions: DoD Continues to Face Challenges of Delayed Delivery of Critical Space Capabilities and Fragmented Leadership," testimony before the Subcommittee on Strategic Forces, Committee on Armed Services, U.S. Senate, May 17, 2017.

Chu, David S. C., and Nurith Berstein, "Decisionmaking for Defense," in Stuart E. Johnson, Martin C. Libicki, and Gregory F. Treverton, eds., *New Challenges New Tools for Defense Decisionmaking*, Santa Monica, Calif.: RAND Corporation, MR-1576-RC, 2003, pp. 13–32. As of November 19, 2019:
https://www.rand.org/pubs/monograph_reports/MR1576.html

Clark, Colin, "Top DoD Official Shank Resigns; SCO Moving to DARPA," *Breaking Defense*, June 17, 2019. As of August 2, 2019:
https://breakingdefense.com/2019/06/top-dod-official-shank-resigns-sco-moving-to-darpa/

Clausen, Robert, Chris Elenbaum, Tim Sejba, and Brendan Hochstein, "Organizing U.S. Space Force Acquisition," briefing, Washington, D.C.: Space Force Planning Task Force Acquisition Planning Team, Headquarters, Air Force, May 9, 2019.

Clodfelter, Mark A., "Molding Airpower Convictions: Development and Legacy of William Mitchell's Strategic Thought," in Phillip S. Meilinger, ed., *The Paths of Heaven: The Evolution of Airpower Theory*, Maxwell Air Force Base, Ala.: Air University Press, 1997, pp. 79–114.

Coast Guard Aviation Association, "United States Coast Guard Transferred to the Department of Transportation," in *United States Coast Guard Aviation History*, Troy, Va., 2017. As of December 3, 2019:
https://cgaviationhistory.org/1967-united-states-coast-guard-transferred-to-the-department-of-transportation/

Cohen, Andrew Wender, "Smuggling, Globalization, and America's Outward State, 1870–1909," *Journal of American History*, Vol. 97, No. 2, September 2010.

Cohen, William S., "The 20-Year Climb to an Elevated CyberCom," *FCW*, October 9, 2017. As of November 15, 2019:
https://fcw.com/articles/2017/10/09/comment-cohen-cyber.aspx

Coleman, Jason, Adam Talaber, and Matthew Woodward, *The Personnel Requirements and Costs of New Military Space Organizations*, Washington, D.C.: Congressional Budget Office, May 2019.

Commission to Assess United States National Security Space Management and Organization, *Report of the Commission to Assess United States National Security Space Management and Organization*, Washington, D.C., 2001. As of August 16, 2019:
https://apps.dtic.mil/dtic/tr/fulltext/u2/a404328.pdf

Committee on Armed Services, House of Representatives, *Declaration of Policy of the United States Concerning National Missile Defense Deployment*, Washington, D.C., 106-39 Part 1, 1999.

Condit, Kenneth W., John H. Johnstone, and Ella W. Nargele, *A Brief History of Headquarters Marine Corps Staff Organization*, Washington, D.C.: Historical Division, Headquarters, Marine Corps, 1971.

Congressional Research Service, *Defense Primer: Special Operations Forces*, Washington, D.C., 2018.

———, *U.S. Special Operations Forces (SOF): Background and Issues for Congress*, Washington, D.C., March 2019.

Correll, John T., "The Air Mail Fiasco," *Air Force Magazine*, March 2008, pp. 60–66.

Cox, Kate, Stephen Jolly, Simon Van Der Staaij, and Christian Van Stolk, *Understanding the Drivers of Organisational Capacity*, Santa Monica, Calif.: RAND Corporation, RR-2189-SAATCHI, 2018. As of September 28, 2019:
https://www.rand.org/pubs/research_reports/RR2189.html

CYBERCOM—*See* U.S. Cyber Command.

Dabrowski, John, *Missile Defense: The First Seventy Years*, Fort Belvoir, Va.: Missile Defense Agency, 2013.

DAF—*See* Department of the Air Force.

Dastyari, Azadeh, *United States Migrant Interdiction and the Detention of Refugees in Guantánamo Bay*, New York: Cambridge University Press, 2015.

Dave, Paresh, "Microsoft Workers Demand It Drop $480 Million U.S. Army Contract," Reuters, February 22, 2019. As of September 28, 2019:
https://www.reuters.com/article/us-microsoft-army/microsoft-workers-demand-it-drop-480-million-u-s-army-contract-idUSKCN1QB2LV?

Davis, Lynn E., Debra Knopman, Michael D. Greenberg, Laurel E. Miller, and Abby Doll, *Choosing a New Organization for Management and Disposition of Commercial and Defense High-Level Radioactive Materials*, Santa Monica, Calif.: RAND Corporation, MG-1230, 2012. As of September 28, 2019:
https://www.rand.org/pubs/monographs/MG1230.html

Davis, Paul K., *Analysis to Inform Defense Planning Despite Austerity*, Santa Monica, Calif.: RAND Corporation, RR-482-OSD, 2014. As of September 28, 2019:
https://www.rand.org/pubs/research_reports/RR482.html

Defense Intelligence Agency, *Challenges to Security in Space*, Washington, D.C., January 2019.

Defense Manpower Data Center, "Active Duty Military Personnel by Service by Rank/Grade," data set, updated September 30, 2019. As of October 2, 2019:
https://www.dmdc.osd.mil/appj/dwp/dwp_reports.jsp

Defense Science Board Permanent Task Force on Nuclear Weapons Surety, *Report on the Unauthorized Movement of Nuclear Weapons*, Washington, D.C.: Office of the Under Secretary of Defense for Acquisition, Technology, and Logistics, February 2008. As of August 28, 2019:
https://www.bits.de/NRANEU/docs/2008-02-Nuclear_Weapons_Surety.pdf

DeJesus, Adam, "Navy Space Cadre and Warfighters Putting Pieces Together," *CHIPS*, June 10, 2014. As of September 28, 2019:
https://www.doncio.navy.mil/chips/ArticleDetails.aspx?ID=5186

Department of Defense, *Responsibility for Development of Space Systems*, Washington, D.C., Department of Defense Directive 5160.32, March 6, 1961.

———, *Quadrennial Defense Review Report*, Washington, D.C., February 6, 2006.

———, *Missile Defense Agency (MDA)*, Washington, D.C., Department of Defense Directive 5134.09, 2009.

———, *Functions of the Department of Defense and Its Major Components*, Washington, D.C., Department of Defense Directive 5100.01, December 21, 2010.

———, *Final Report on Organizational and Management Structure for the National Security Space Components of the Department of Defense*, Washington, D.C., August 9, 2018. As of August 1, 2019:
https://media.defense.gov/2018/Aug/09/2001952764/-1/-1/1/ORGANIZATIONAL-MANAGEMENT-STRUCTURE-DOD-NATIONAL-SECURITY-SPACE-COMPONENTS.PDF

———, *United States Space Force*, Washington, D.C., February 2019a. As of August 1, 2019:
https://media.defense.gov/2019/Mar/01/2002095012/-1/-1/1/UNITED-STATES-SPACE-FORCE-STRATEGIC-OVERVIEW.PDF

———, proposed amendments to U.S. Code to stand up U.S. Space Force, March 2019b. As of December 12, 2019:
https://media.defense.gov/2019/Mar/01/2002095010/-1/-1/1/UNITED-STATES-SPACE-FORCE-LEGISLATIVE-PROPOSAL.PDF

Department of Homeland Security, *Department of Homeland Security: U.S. Coast Guard Budget Overview Fiscal Year 2020 Congressional Justification*, Washington, D.C., 2019.

Department of the Air Force, *Air Force Global Strike Command Strategic Plan 2018*, Washington, D.C., 2018.

———, *Comprehensive Plan for the Organizational Structure of the U.S. Space Force*, Washington, D.C., February 2020.

Department of the Navy, Office of Budget, *Highlight of the Department of the Navy FY2019 Budget*, Washington, D.C., 2018.

DeVito, Allison, "Space Force Legislative Proposal," briefing, Washington, D.C.: Office of the Secretary of the Air Force, Deputy General Counsel for Acquisition, February 27, 2019.

Dewar, James A., *Assumption-Based Planning: A Tool for Reducing Avoidable Surprises*, New York: Cambridge University Press, 2002.

Dewar, James A., Carl H. Builder, William M. Hix, and Morlie Levin, *Assumption-Based Planning: A Planning Tool for Very Uncertain Times*, Santa Monica, Calif.: RAND Corporation, MR-114-A, 1993. As of March 16, 2019:
https://www.rand.org/pubs/monograph_reports/MR114.html

de Zwart, Melissa, "It's Not Clear Where Trump's 'Space Force' Fits Within International Agreement on Peaceful Use of Space," *Conversation*, June 19, 2019.

DiPane, James, "Should Cyber Command and the NSA Have Separate Leadership? How to Decide," Heritage Foundation, May 2, 2019.

DoD—*See* Department of Defense.

Dolman, Everett C., *Astropolitik: Classical Geopolitics in the Space Age*, New York: Routledge, 2001.

Donovan, Matthew, "Unleashing the Power of Space: The Case for a Separate US Space Force," *War on the Rocks*, August 1, 2019. As of October 1, 2019:
https://warontherocks.com/2019/08/unleashing-the-power-of-space-the-case-for-a-separate-u-s-space-force/

Dudley, Ian, "Highway to Space," Air Force Space Command, January 15, 2016. As of September 28, 2019:
https://www.afspc.af.mil/News/Features/Display/Article/731636/highway-to-space/

Epstein, Marc J., *Making Sustainability Work: Best Practices in Managing and Measuring Corporate Social, Environmental and Economic Impacts*, London: Routledge, 2008.

Erwin, Sandra, "Shanahan Officially Establishes the Space Development Agency," *Space News*, March 13, 2019.

Faber, Peter R., "Interwar US Army Aviation and the Air Corps Tactical School: Incubators of American Airpower," in Phillip S. Meilinger, ed., *The Paths of Heaven: The Evolution of Airpower Theory*, Maxwell Air Force Base, Ala.: Air University Press, 1997.

Faulkenberry, Matthew E., *Critical Review of the Navy Space Cadre*, thesis, Monterey, Calif.: Naval Postgraduate School, June 2014. As of September 20, 2019:
https://apps.dtic.mil/dtic/tr/fulltext/u2/a607753.pdf

Ferdinando, Lisa, "Cybercom to Elevate to Combatant Command," *DoD News*, May 3, 2018. As of September 20, 2019:
https://www.jcs.mil/Media/News/News-Display/Article/1512442/cybercom-to-elevate-to-combatant-command/

Finney, Robert E., *History of the Air Corps Tactical School, 1920–1940*, Washington, D.C.: Air Force History and Museums Program, United States Air Force, 1998.

Futrell, Robert Frank, *Ideas, Concepts, and Doctrine: Basic Thinking in the United States Air Force 1907–1960*, Maxwell Air Force Base, Ala.: Air University Press, 1989.

Galer, John, "Only a Separate Service Can Create a Space Culture," *SpaceNews*, June 9, 2019. As of September 30, 2019:
https://spacenews.com/op-ed-only-a-separate-service-can-create-a-space-culture/

Galvin, Tom, ed., *Defense Management: Primer for Senior Leaders*, 1st ed., Carlisle, Pa.: U.S. Army War College, 2018.

Gansler, Jacques S., and William Lucyshyn, *Improving the DoD's Tooth-to Tail Ratio*, College Park: Center for Public Policy and Private Expertise, University of Maryland, February 2014.

Garamone, Jim, "Air Force Global Strike Command Will Stress Nuclear Mission," American Forces Press Service, August 7, 2009. As of August 28, 2019:
https://www.malmstrom.af.mil/News/Article-Display/Article/348323/
air-force-global-strike-command-will-stress-nuclear-mission/

Gates, Robert M., *Duty: Memoirs of a Secretary at War*, New York: Vintage, 2015.

Gentile, Gian P., *How Effective Is Strategic Bombing? Lessons Learned from World War II and Kosovo*, New York: New York University Press, 2001.

George, Justin Paul, "History of Anti-Satellite Weapons: US Tested 1st ASAT Missile 60 Years Ago," *The Week*, March 27, 2019.

Gohd, Chelsea, "Peace in Space: World Leaders Discuss Peaceful Uses of the Cosmos at UN Event," Space.com, September 22, 2018. As of July 21, 2019:
https://www.space.com/41907-space-trust-peace-in-space.html

Gompert, David C., and Jeffrey A. Isaacson, *Planning a Ballistic Missile Defense System of Systems: An Adaptive Strategy*, Santa Monica, Calif.: RAND Corporation, IP-181, 1999. As of September 20, 2019:
https://www.rand.org/pubs/issue_papers/IP181.html

Government Accountability Office, *Defense Cybersecurity: DoD's Monitoring of Implementing Cyber Strategies Can Be Strengthened*, Washington, D.C., August 2017. As of September 20, 2019:
https://www.gao.gov/products/GAO-17-512

———, *DoD Training: U.S. Cyber Command and Services Should Take Actions to Maintain a Trained Cyber Mission Force*, Washington, D.C., March 2019. As of September 20, 2019:
https://www.gao.gov/assets/700/697268.pdf

Greer, Thomas H., *The Development of Air Doctrine in the Army Air Arm: 1917–1941*, Washington, D.C.: Office of Air Force History, U.S. Air Force, 1985.

Gruss, Mike, and Aaron Mehta, "Space Force to Cost $2 Billion, Include 15,000 Personnel in First Five Years," *Defense News*, March 1, 2019.

Haasnoot, Marjolijn, Jan H. Kwakkel, Warren E. Walker, and Judith ter Maat, "Dynamic Adaptive Policy Pathways: A Method for Crafting Robust Decisions for a Deeply Uncertain World," *Global Environmental Change*, Vol. 23, 2013, pp. 485–498.

Hall, R. Cargill, *Military Space and National Policy: Record and Interpretation*, Arlington, Va.: George C. Marshall Institute, 2006.

Harrison, Todd, *How Much Will the Space Force Cost?* Washington, D.C.: Center for Strategic and International Studies, November 2018a.

———, "A Space Force Is Worth the Price," *The Hill*, November 27, 2018b.

Harrison, Todd, Kaitlyn Johnson, and Thomas G. Roberts, *Space Threat Assessment 2018*, Washington, D.C.: Center for Strategic and International Studies, April 2018.

———, *Space Threat Assessment 2019*, Washington, D.C.: Center for Strategic and International Studies, April 2019.

Harwell, Drew, "Google to Drop Pentagon AI Contract After Employee Objections to the 'Business of War,'" *Washington Post*, June 1, 2018. As of September 20, 2019:
https://www.washingtonpost.com/news/the-switch/wp/2018/06/01/google-to-drop-pentagon-ai-contract-after-employees-called-it-the-business-of-war/?utm_term=.12ee1056fac2

Headquarters, Department of the Army, *Army Space Operations*, Washington, D.C., Field Manual 3-14, Change 2, February 13, 2018.

Helms, Christian P., *The Digital GCC: USCYBERCOM as a Combatant Command*, Maxwell Air Force Base, Ala.: Air Command and Staff College, Air University, April 2015. As of September 20, 2019:
https://apps.dtic.mil/dtic/tr/fulltext/u2/1012758.pdf

Hoar, Margaux, Eileen Chollet, Kory Fierstine, Chao Pan, and Mark Rosen, *Plan for a United States Department of the Space Force: Executive Summary*, Arlington Va.: Center for Naval Analysis, November 2018.

Hunt, Cameron S., "AFOTEC Maintains Mission Readiness Through Operational Testing," U.S. Air Force, February 26, 2019. As of November 25, 2019:
https://www.af.mil/News/Article-Display/Article/1767944/
afotec-maintains-mission-readiness-through-operational-testing/

Insinna, Valerie, "Trump Officially Organizes the Space Force Under the Air Force . . . for Now," *Defense News*, February 19, 2019.

Joint Chiefs of Staff, *The National Military Strategy of the United States of America*, Washington, D.C., 2004.

———, *Space Operations*, Washington, D.C., Joint Publication 3-14, April 10, 2018a.

———, *Cyberspace Operations*, Washington, D.C., Joint Publication 3-12, June 8, 2018b.

Judson, Jen, "Where Do Special Forces Fit in the National Defense Strategy?" *Defense News*, May 17, 2019.

Kanter, Arnold, *Defense Politics: A Budgetary Perspective*, Chicago: University of Chicago Press, 1979.

Kaplan, Robert S., and David P. Norton, "The Balanced Scorecard: Measures That Drive Performance," *Harvard Business Review*, January–February 1992, pp. 71–79.

———, *The Balanced Scorecard: Translating Strategy into Action*, Boston: Harvard Business School Press, 1996.

———, *The Strategy-Focused Organization: How Balanced Scorecard Companies Thrive in the New Business Environment*, Boston: Harvard Business School Press, 2000.

Keaney, Thomas A., and Eliot A. Cohen, *Revolution in Warfare? Air Power in the Persian Gulf War*, Annapolis, Md.: Naval Institute Press, 1995.

Keiser, Gordon W., *The US Marine Corps and Defense Unification 1944–47: The Politics of Survival*, Washington, D.C.: National Defense University Press, 1982.

Kelman, Steven, *Unleashing Change: A Study of Organizational Renewal in Government*, Washington, D.C.: Brookings Institution Press, 2005.

Kennett, Lee, *The First Air War: 1914–1918*, New York: The Free Press, 1991.

Keyte, Clive, *How to Develop Meaningful Key Performance Indicators: A Step by Step Guide*, Winchester, UK: Intrafocus Limited, 2018. As of August 31, 2019:
https://static.intrafocus.com/uploads/2018/05/How-to-Develop-Meaningful-Key-Performance-Indicators-V7-web.pdf

King, Irving H., *The Coast Guard Under Sail: The U.S. Revenue Cutter Service, 1789–1865*, Annapolis, Md.: Naval Institute Press, 1989.

Korb, Lawrence J., Sean Duggan, and Laura Conley, *Building a U.S. Coast Guard for the 21st Century*, Washington, D.C.: Center for American Progress, June 2010.

Kotter, John P., "Winning at Change," *Leader to Leader*, Vol. 10, Fall 1998, pp. 27–33.

Kowalski, James M., *Air Force Global Strike Command*, briefing, Barksdale Air Force Base, La., May 2013.

Kwakkel, Jan, "Managing Deep Uncertainty: Exploratory Modeling, Adaptive Plans, and Joint Sense Making," *Integration and Implementation Insights*, August 15, 2017. As of August 4, 2019: https://i2insights.org/2017/08/15/managing-deep-uncertainty/

Lange, Katie, "Cybercom Becomes DoD's 10th Unified Combatant Command," *DoDLive*, May 3, 2018. As of November 15, 2019: http://www.dodlive.mil/2018/05/03/cybercom-to-become-dods-10th-unified-combatant-command/

Laurie, Clayton D., *Congress and the National Reconnaissance Office*, Chantilly, Va.: Office of the Historian, National Reconnaissance Office, June 2001.

Lay, James S., "Memorandum for the National Security Council: U.S. Policy on Continental Defense," Washington, D.C.: White House, 1960.

Lempert, Robert J., Steven W. Popper, and Steven C. Bankes, "Confronting Surprise," *Social Science Computer Review*, Vol. 20, No. 4, Winter 2002, pp. 420–440.

Lempert, Robert J., Horacio R. Trujillo, David Aaron, James A. Dewar, Sandra H. Berry, and Steven W. Popper, *Comparing Alternative U.S. Counterterrorism Strategies: Can Assumption-Based Planning Help Elevate the Debate?* Santa Monica, Calif.: RAND Corporation, DB-548-RC, 2008. As of November 15, 2019: https://www.rand.org/pubs/documented_briefings/DB548.html

Light, Paul C., *The Four Pillars of High Performance: How Robust Organizations Achieve Extraordinary Results*, New York: McGraw-Hill, 2005.

Londoño, Ernesto, "US Air Force Relieves Nine Officers Following Nuclear Cheating Probe," *Washington Post*, March 27, 2014.

Loverro, Douglas, "Why the United States Needs a Space Force," *SpaceNews*, June 25, 2018.

Lupton, David E., *On Space Warfare: A Space Power Doctrine*, Maxwell Air Force Base, Ala.: Air University Press, 1998.

Lutes, Charles D., and Peter L. Hays, eds., *Toward a Theory of Spacepower: Selected Essays*, Washington, D.C.: National Defense University Press, 2011.

Lynch, Justin, "Inside the Pentagon's Struggle to Build a Cyber Force," *Fifth Domain*, October 18, 2018a. As of November 15, 2019: https://www.fifthdomain.com/dod/2018/10/29/inside-the-pentagons-struggle-to-build-a-cyber-force/

———, "Why Recruiting Cyberwarriors in the Military Is Harder Than Retaining Forces," *Fifth Domain*, November 1, 2018b. As of November 15, 2019: https://www.fifthdomain.com/dod/2018/11/01/why-recruiting-cyber-warriors-in-the-military-is-harder-than-retaining-forces/

MacDougall, Walter A., *The Heavens and the Earth: A Political History of the Space Age*, Baltimore, Md.: Johns Hopkins University Press, 1985.

Malboussin, Michael J., "The True Measures of Success," *Harvard Business Review*, October 2012. As of August 31, 2019:
https://hbr.org/2012/10/the-true-measures-of-success

Marchau, Vincent A. W. J., Warren E. Walker, Pieter J. T. M. Bloemen, and Steven W. Popper, eds., *Decision Making Under Deep Uncertainty: From Theory to Practice*, Cham, Switzerland: Springer, 2019.

McDermott, Roger N., *Russia's Strategic Mobility: Supporting Hard Power to 2020?* Stockholm: Swedish Defence Research Agency, April 2013.

McGrath, John J., *The Other End of the Spear: The Tooth-to-Tail Ratio (T3R) in Modern Military Operations*, Fort Leavenworth, Kan.: Combat Studies Institute Press, 2007.

McInnis, Kathleen J., and Stephen M. McCall, *"Space Force" and Related DoD Proposals: Issues for Congress*, Washington, D.C.: Congressional Research Service, April 8, 2019.

MDA—*See* Missile Defense Agency.

Meilinger, Phillip S., "When the Fortress Went Down," *Air Force Magazine*, October 2004, pp. 78–82.

Mercurio, Nicholas, "New Squadron, 614th CTS, Activates at Vandenberg," Vandenberg Air Force Base, Calif., December 21, 2016. As of September 28, 2019:
https://www.vandenberg.af.mil/News/Article-Display/Article/1036059/new-squadron-614th-cts-activates-at-vandenberg/

Missile Defense Agency, "Integrated Flight Tests at U.S. Army Kwajalein Atoll/Ronald Reagan Ballistic Missile Defense Test Site (USAKA/RTS) Environmental Assessment," August 12, 2012. As of September 28, 2019:
https://www.mda.mil/global/documents/pdf/env_IFT_USAKA-RTS_FinalFONSI.pdf

———, "Space Tracking and Surveillance System," fact sheet, 17-MDA-9108, March 27, 2017. As of June, 2019:
https://www.mda.mil/global/documents/pdf/stss.pdf

———, *Fiscal Year (FY) 2019 Budget Estimates: Overview*, Fort Belvoir, Va., 2018.

———, "Elements: Ground-Based Midcourse Defense (GMD)," webpage, last updated October 3, 2019a. As of December 3, 2019:
https://www.mda.mil/system/gmd.html

———, *Notification and Federal Employee Antidiscrimination and Retaliation Act (NoFEAR): Fiscal Year 2018 Report*, Washington, D.C., 2019b.

Mitchell, William, *Our Air Force: The Key to National Defense*, New York: Dutton, 1921.

———, *Winged Defense*, New York: P.G. Putnam's Sons, 1925.

———, *Skyways*, Philadelphia: J.B. Lippincott Company, 1930.

Modley, Rudolf, and Thomas J. Cawley, *Aviation Facts and Figures, 1953*, Washington, D.C.: Lincoln Press, 1953.

Morgan, Forrest E., "The Concept of Airpower: Its Emergence, Evolution, and Future," in Richard B. Hallion, Roger Cliff, and Phillip C. Saunders, eds., *The Chinese Air Force: Evolving Concepts, Roles, and Capabilities*, Washington, D.C.: National Defense University Press, 2012, pp. 1–31.

Nakashima, Ellen, and Aaron Gregg, "NSA Is Losing Talent over Low Pay, Flagging Morale and Reorganization," *Chicago Tribune*, January 2, 2018. As of September 28, 2019:
https://www.chicagotribune.com/news/nationworld/ct-nsa-losing-talent-20180102-story.html

Nalty, Bernard C., "Reaction to the War in Europe," in Bernard C. Nalty, ed., *Winged Shield, Winged Sword: A History of the United States Air Force*, Vol. 1, Washington, D.C.: Air Force History and Museums Program, U.S. Air Force, 1997, pp. 165–200.

National Reconnaissance Office, "About the NRO: Who We Are," webpage, undated. As of October 2, 2019:
https://www.nro.gov/About-NRO/The-National-Reconnaissance-Office/Who-We-Are/

National Security Agency, "NSA Welcomes General Paul Nakasone as Agency Director," May 4, 2018. As of October 2, 2019:
https://www.nsa.gov/News-Features/News-Stories/Article-View/Article/1623070/
nsa-welcomes-general-paul-nakasone-as-agency-director/

National Security Council, *U.S. Policy on Continental Defense*, Washington, D.C., NSC 5802, 1958.

National Security Space Institute, "About," webpage, undated-a. As of September 28, 2019:
https://www2.peterson.af.mil/nssi/

———, "Partners," webpage, undated-b. As of September 20, 2019:
https://www2.peterson.af.mil/nssi/

Naval Research Laboratory, "Blossom Point Tracking Facility (BPTF)," fact sheet, April 2019. As of September 28, 2019:
https://www.nrl.navy.mil/space/system/files/resources/
Blossom%20Point%20Tracking%20Facility.pdf

Nellis Air Force Base, "United States Air Force Weapons School," fact sheet, May 10, 2016. As of October 2, 2019:
https://www.nellis.af.mil/About/Fact-Sheets/Display/Article/284156/
united-states-air-force-weapons-school/

NRL—*See* Naval Research Laboratory.

NSA—*See* National Security Agency.

NSSI—*See* National Security Space Institute.

Oberg, James E., *Space Power Theory*, Colorado Springs, Colo.: U.S. Air Force Academy, 1999.

Ochmanek, David, *Improving Force Development Within the Department of Defense: Diagnosis and Potential Prescriptions*, Santa Monica, Calif.: RAND Corporation, PE-302-RC, July 2018. As of October 2, 2019:
https://www.rand.org/pubs/perspectives/PE302.html

Ochmanek, David A., and Lowell Schwartz, *The Challenge of Nuclear-Armed Regional Adversaries*, Santa Monica, Calif.: RAND Corporation, MG-671-AF, 2008. As of December 3, 2019:
https://www.rand.org/pubs/monographs/MG671.html

O'Connell, Aaron B., *Underdogs: The Making of the Modern Marine Corps*, Cambridge, Mass.: Harvard University Press, 2012.

Office of Statistical Control, U.S. Army Air Forces, *Army Air Forces Statistical Digest, World War II*, Washington, D.C., December 1945.

Office of the Chairman of the Joint Chiefs of Staff, *DoD Dictionary of Military and Associated Terms*, Washington, D.C., October 2019.

Ostroff, Frank, "Change Management in Government," *Harvard Business Review*, May 2006.

Palmer, Gary W., "Guarding the Coast: Alien Migrant Interdiction Operations at Sea," *Connecticut Law Review*, Vol. 29, 1996, pp. 158–179.

Panda, Ankit, "Russia Conducts New Test of 'Nudol' Anti-Satellite System," *The Diplomat*, April 2, 2018.

Pape, Robert A., *Bombing to Win: Air Power and Coercion in War*, Ithaca, N.Y.: Cornell University Press, 1996.

Parker, William E., IV, *Cyber Workforce Retention*, Maxwell Air Force Base, Ala.: Air Force Research Institute, Air University, October 2016. As of November 15, 2019:
https://media.defense.gov/2017/May/11/2001745609/-1/-1/0/CPP_0002_PARKER_CYBER_WORKFORCE_RETENTION.PDF

Parmenter, David, *Key Performance Indicators: Developing, Implementing, and Using Winning KPIs*, New York: Wiley & Sons, 2015.

Paul, Christopher, Isaac R. Porche III, and Elliot Axelband, *The Other Quiet Professionals: Lessons for Future Cyber Forces from the Evolution of Special Forces*, Santa Monica, Calif.: RAND Corporation, RR-780-A, 2014. As of November 15, 2019:
https://www.rand.org/pubs/research_reports/RR780.html

Peebles, Curtis, *High Frontier: The United States Air Force and the Military Space Program*, Washington, D.C.: Air Force History and Museums Program, 1997.

Pickrell, Ryan, "Chinese Hackers Are Reportedly Stealing Loads of US Navy Secrets, and the Navy Is Scrambling to Stop It," *Business Insider*, December 14, 2018. As of November 15, 2019:
https://www.businessinsider.com/chinese-hackers-us-navy-secrets-2018-12

Pomerleau, Mark, "Rogers: Cyber Doesn't Needs Its Own Military Branch," *Defense Systems*, January 21, 2016. As of November 15, 2019:
https://defensesystems.com/articles/2016/01/21/rogers-cyber-doesnt-need-to-be-separate-branch.aspx

Price, Dave, "JTAGS: Army Theater Missile Warning," *Air Defense Artillery*, October–December 2005, pp. 16–18. As of September 28, 2019:
https://sill-www.army.mil/ada-online/pb-44/_docs/2005/10-12/Oct-Dec%202005.pdf

Public Law 219, An Act to Integrate Certain Personnel of the Former Bureau of Marine Inspection and Navigation and the Bureau of Customs into the Regular Coast Guard, to Establish the Permanent Commissioned Personnel Strength of the Coast Guard, and for Other Purposes, July 23, 1947

Public Law 45-263, Posse Comitatus Act of 1878, June 18, 1878.

Public Law 66-66, National Prohibition Act, October 28, 1919.

Public Law 80-253, The National Security Act of 1947, July 26, 1947.

Public Law 85-911, Boating Act of 1958, September 2, 1958.

Public Law 92-75, Federal Boat Safety Act of 1971, August 10, 1971.

Public Law 98-473, The National Narcotics Act of 1984, October 12, 1984.

Public Law 99-433, Barry Goldwater Department of Defense Reorganization Act of 1986, October 1, 1986.

Public Law 99-661, National Defense Authorization Act for Fiscal Year 1987, November 14, 1986.

Public Law 106-38, National Missile Defense Act of 1999, July 22, 1999.

Public Law 107-295, Maritime Transportation Security Act of 2002, November 25, 2002.

Public Law 115-232, John S. McCain National Defense Authorization Act for Fiscal Year 2019, Title V, Military Personnel Policy; Subtitle A, Officer Personnel Policy, Sections 501–502, August 13, 2018.

Public Law 116-92, National Defense Authorization Act for Fiscal Year 2020, Subtitle D, United States Space Force, Sections 951–961, December 20, 2019.

Ramsby, Corey M., and Panayotis A. Yannakogeorgos, "A Reality Check on the Cyber Force," *Strategic Studies Quarterly*, Summer 2016, pp. 128–129.

Reagan, Ronald, "Address to the Nation on Defense and National Security," Washington, D.C., March 23, 1983.

Rice, Condoleezza, *Missile Defense Papers*, Washington, D.C.: White House, 2001.

Richard, Pierre J., Timothy M. Devinney, George S. Yip, and Gerry Johnson, "Measuring Organizational Performance: Towards Methodological Best Practice," *Journal of Management*, Vol. 35, No. 3, 2009, pp. 718–804.

Ricks, Thomas E., and Joby Warrick, "Tough Punishment Expected for Warhead Errors; Officers May Lose Commands After Nuclear Missiles Were Flown on Bomber," *Washington Post*, October 18, 2007.

Ross, Shirley M., *Final Report: Nuclear Enterprise Human Capital Strategy and Execution Progress 2010–2012*, Washington, D.C.: Deputy Chief of Staff of the Air Force for Manpower, Personnel, and Services, 2012.

Royx, Barry, *Space Cadre Quick Look*, 3rd ed., Peterson Air Force Base, Colo.: Air Force Space Command, February 2019, internal document.

Rumsfeld, Donald, "Intelligence Side Letter to the Rumsfeld Commission Report," March 18, 1999. As of December 3, 2019:
https://fas.org/irp/threat/missile/sideletter.htm

———, *Missile Defense Program Direction*, Washington, D.C.: Department of Defense, 2002.

"S-500 or A-235? Russia Tests New Missile Defence System with Extreme Range," *Military Watch*, June 4, 2019.

Schriever Air Force Base, "Air Force Technical Applications Center, Det. 46," fact sheet, July 2016. As of November 15, 2019:
https://www.schriever.af.mil/About-Us/Fact-Sheets/Air-Force-Technical-Applications-Center-Det-46/

———, "25th Space Range Squadron," fact sheet, June 22, 2018. As of September 28, 2019:
https://www.schriever.af.mil/About-Us/Fact-Sheets/Display/Article/1558204/25th-space-range-squadron/

Secretary of Defense, "Memorandum: U.S. Space Force Planning Team," Washington, D.C.: Office of the Secretary of Defense, February 21, 2019.

Secretary of the Air Force, "Memorandum: Establishment of the US Space Force Planning Task Force," Washington, D.C.: U.S. Air Force, February 22, 2019.

Seligman, Lara, "Space Force Is Trump's Answer to New Russian and Chinese Weapons," *Foreign Policy*, August 10, 2019.

Senate Armed Services Committee, *FY20 National Defense Authorization Act Executive Summary*, Washington, D.C., 2019.

Sherman, Jason, "MDA Identifies Nine Companies Competing for Space Sensor Layer," *Inside Defense*, March 7, 2019. As of September 28, 2019:
https://insidedefense.com/daily-news/mda-identifies-nine-companies-competing-space-sensor-layer

Shiner, John F., "The Coming of the GHQ Air Force, 1925–1935," in Bernard C. Nalty, ed., *Winged Shield, Winged Sword: A History of the United States Air Force*, Vol. 1, Washington, D.C.: Air Force History and Museums Program, U.S. Air Force, 1997a, pp. 101–134.

———, "The Heyday of the GHQ Air Force, 1935–1939," in Bernard C. Nalty, ed., *Winged Shield, Winged Sword: A History of the United States Air Force*, Vol. 1, Washington, D.C.: Air Force History and Museums Program, U.S. Air Force, 1997b, pp. 135–164.

SOCOM—*See* U.S. Special Operations Command.

Spoehr, Thomas, and James Di Pane, "Elevating Cyber Command: An Overdue Step Towards Enhancing Military Cyber Operations," Heritage Foundation, October 1, 2018. As of November 15, 2019:
https://www.heritage.org/cybersecurity/commentary/
elevating-cyber-command-overdue-step-towards-enhancing-military-cyber

Stavridis, James, "The U.S. Needs a Cyber Force More Than a Space Force," *Bloomberg*, August 14, 2018.

Stavridis, James, and David Weinstein, "Time for a U.S. Cyber Force," *Proceedings*, Vol. 140, January 2014. As of November 15, 2019:
https://www.usni.org/magazines/proceedings/2014/january/time-us-cyber-force

Stuart, Douglas T., ed., *Organizing for National Security*, Carlisle, Pa.: Strategic Studies Institute, November 2000.

Swarts, Phillip, "Rogers Calls for Separate 'Space Corps' Within the Air Force," *Space News*, April 4, 2017.

Tate, James P., *The Army and Its Air Corps: 1919–1941*, Maxwell Air Force Base, Ala.: Air University Press, 1998.

Thiesen, William H., "The Coast Guard's World War II Crucible," *Naval History Magazine*, Vol. 30, No. 5, October 2016.

Thornhill, Paula G., *The Crisis Within: America's Military and the Struggle Between the Overseas and Guardian Paradigms*, Santa Monica, Calif.: RAND Corporation, RR-1420-AF, 2016. As of November 15, 2019:
https://www.rand.org/pubs/research_reports/RR1420.html

———, *Demystifying the American Military: Institutions, Evolution, and Challenges Since 1789*, Annapolis, Md.: Naval Institute Press, 2019.

U.S. Army Space and Missile Defense Command, "1st Space Brigade," webpage, undated. As of September 28, 2019:
https://www.smdc.army.mil/ORGANIZATION/1st-Space-Brigade/

———, "Ronald Reagan Ballistic Missile Defense Test Site at Kwajalein Atoll (RTS)," fact sheet, 2019. As of September 28, 2019:
https://www.smdc.army.mil/Portals/38/Documents/Publications/Fact_Sheets/
RTS.pdf?ver=2019-01-08-143444-290

U.S. Army Space and Missile Defense Command and Army Forces Strategic Command, "RSSC: Regional Satellite Communications Support Centers," fact sheet, undated. As of September 28, 2019:
https://www.smdc.army.mil/Portals/38/Documents/Publications/Fact_Sheets/RSSC.pdf

U.S. Coast Guard, *Doctrine for the U.S. Coast Guard*, Washington, D.C., Coast Guard Publication 1, February 2014.

———, *U.S. Coast Guard Missions: A Historical Timeline*, Washington, D.C., June 25, 2017. As of December 3, 2019:
https://media.defense.gov/2017/Jun/25/2001768442/-1/-1/0/USCGMISSIONSTIMELINE.PDF

———, "U.S. Coast Guard Fact Sheet: Fiscal Year 2019 President's Budget," 2018.

U.S. Code, Title 10, Armed Forces. As of November 18, 2019:
https://www.govinfo.gov/app/details/USCODE-2018-title10/

U.S. Code, Title 10, Armed Forces; Section 101, Definitions. As of November 18, 2019:
https://www.govinfo.gov/app/details/USCODE-2018-title10/
USCODE-2018-title10-subtitleA-partI-chap1-sec101

U.S. Code, Title 10, Armed Forces; Subtitle A, General Military Law; Part I, Organization and General Military Powers; Chapter 6, Combatant Commands; Section 167, Unified Combatant Command for Special Operations Forces. As of December 3, 2019:
https://www.govinfo.gov/app/details/USCODE-2011-title10/
USCODE-2011-title10-subtitleA-partI-chap6-sec167

U.S. Code, Title 10, Armed Forces; Subtitle C, Navy and Marine Corps; Part I, Organization; Chapter 507, Composition of the Department of the Navy; Section 5063, United States Marine Corps: Composition; Functions. As of December 3, 2019:
https://www.govinfo.gov/app/details/USCODE-2010-title10/
USCODE-2010-title10-subtitleC-partI-chap507-sec5063

U.S. Code, Title 50, War and National Defense. As of November 15, 2019:
https://www.govinfo.gov/app/details/USCODE-2017-title50/USCODE-2017-title50-front

U.S. Cyber Command, "Mission and Vision," webpage, undated-a. As of November 15, 2019:
https://www.cybercom.mil/About/Mission-and-Vision/

———, "U.S. Cyber Command History," webpage, undated-b. As of November 15, 2019:
https://www.cybercom.mil/About/History/

———, *Achieve and Maintain Cyberspace Superiority: Command Vision for US Cyber Command*, Fort Meade, Md., April 2018. As of November 15, 2019:
https://www.cybercom.mil/Portals/56/Documents/USCYBERCOM%20Vision%20April%202018.pdf

U.S. Cyber Command Public Affairs, "Cyber Mission Force Achieves Full Operational Capability," U.S. Cyber Command, May 17, 2018.

U.S. Senate Armed Services Committee, statement of General Paul M. Nakasone, Commander, United States Cyber Command, before the Senate Committee on Armed Services, Washington, D.C., February 14, 2019.

U.S. Special Operations Command, History and Research Office, *United States Special Operations Command History: 1987–2007*, Tampa, Fla., 2007.

Vandenberg Air Force Base, "381st Training Group," fact sheet, December 6, 2016. As of November 15, 2019:
https://www.vandenberg.af.mil/About-Us/Fact-Sheets/Display/Article/1021247/
?3%2081st-training-group=)/

Vasquez, Arielle, "527th SAS: Preparing Warfighters Now, into the Future," Air Force Space Command, February 7, 2018. As of November 15, 2019:
https://www.afspc.af.mil/News/Article-Display/
Article/1436085/527th-sas-preparing-warfighters-now-into-the-future/

Warren, Drake, Heather Krull, Jennifer Lamping Lewis, Aisha Najera Chesler, Ellen M. Pint, and J. Michael Gilmore, Assessment of Alternative Funding Models for Activities in RDECOM (Now CCDC) and ATEC, Santa Monica, Calif.: RAND Corporation, RR-2818-A, 2020. As of February 10, 2020:
https://www.rand.org/pubs/research_reports/RR2818.html

Weeden, Brian, Through a Glass, Darkly: Chinese, American, and Russian Anti-Satellite Testing in Space, Washington, D.C.: Secure World Foundation, March 17, 2014.

Weeden, Brian, and Victoria Samson, eds., Global Counterspace Capabilities: An Open Source Assessment, Washington, D.C.: Secure World Foundation, April 2018.

White, Dottie K., "U.S. Army Space and Missile Defense School House Earns 'Institute of Excellence' from TRADOC," U.S. Army, November 15, 2016. As of August 30, 2019:
https://www.army.mil/article/178342/us_army_space_and_missile_defense_school_house_earns_institute_of_excellence_from_tradoc

White House, "Interdepartmental Board on International Service of Ice Observation, Ice Patrol, and Ocean Derelict Destruction," Washington, D.C., Executive Order No. 2458, September 20, 1916.

———, Intelligence Capabilities: 1992–2005, Washington, D.C., National Security Directive 67, March 30, 1992.

———, "Presidential Memorandum for the Secretary of Defense," Washington, D.C., August 18, 2017. As of November 15, 2019:
https://www.whitehouse.gov/presidential-actions/presidential-memorandum-secretary-defense/

———, National Cyber Strategy of the United States of America, Washington, D.C., September 2018.

———, Establishment of the United States Space Force, Washington, D.C., Space Policy Directive–4, February 19, 2019. As of August 14, 2019:
https://www.whitehouse.gov/presidential-actions/
text-space-policy-directive-4-establishment-united-states-space-force/

Whittington, Michael C., A Separate Space Force: An 80-Year-Old-Argument, Montgomery, Ala.: Air War College, Maxwell Paper No. 20, May 2000.

Young, A. Thomas, Edward Anderson, Lyle Bien, Ronald R. Fogleman, Keith Hall, Lester Lyles, and Hans Mark, Leadership, Management, and Organization for National Security Space: Report to Congress on the Organization and Management of National Security Space, Alexandria, Va.: Institute for Defense Analyses, July 2008. As of August 16, 2019:
https://spacepolicyonline.com/pages/images/stories/Allard_Commission_Report.pdf

Zimmerman, S. Rebecca, Kimberly Jackson, Natasha Lander, Colin Roberts, Dan Madden, and Rebeca Orie, Movement and Maneuver: Culture and the Competition for Influence Among the U.S. Military Services, Santa Monica, Calif.: RAND Corporation, RR-2270-OSD, 2019. As of August 1, 2019:
https://www.rand.org/pubs/research_reports/RR2270.html